Karen Hemmings

Occupational Health:
A Practical Guide for Managers

Occupational Health: A Practical Guide for Managers examines the reasons why employers should address the health of their workforce. It explores the implications that health issues have for bottom line performance and offers guidelines for developing an effective occupational health policy and associated procedures.

The authors are both senior professionals in the field and have a wealth of industrial experience to draw on. They have used their first-hand knowledge to produce a practical introduction to occupational health which includes the latest legislative requirements and current recommended practice. Material is presented in an easy-to-read, non-technical format, making the book particularly useful to the reader with no previous knowledge of the field. Short executive summaries at the beginning of each chapter highlight key action points for quick reference, while a more extensive reference section at the end of the book signposts the way to detailed information on specific issues.

The book is in four parts. Part I provides an introduction to the history and scope of occupational health, discusses contemporary practice and looks at the legal and organisational factors involved. Part II examines some of the most common issues in occupational health, including stress management, anti-smoking policies and health promotion. Specific considerations surrounding the employment of women and people with disabilities are dealt with in Part III. The final part offers a practical approach to two key aspects of health and safety legislation: regulations relating to display screen equipment and the manual handling of loads.

Occupational Health: A Practical Guide for Managers is essential reading for occupational health practitioners and all professionals involved in making and implementing health and safety policy. It is also an extremely useful resource for students seeking to gain an understanding of health issues in the workplace.

Dr Ann Fingret is Senior Consultant in Occupational Medicine at the Royal Marsden NHS Trust. **Alan Smith**, MBE, is Chairman of the National Advisory Council on the Employment of People with Disabilities. Both have considerable experience of working within various fields of industry.

Occupational Health:
A Practical Guide for Managers

Ann Fingret and Alan Smith

London and New York

First published 1995
by Routledge
11 New Fetter Lane, London EC4P 4EE

Simultaneously published in the USA and Canada
by Routledge
29 West 35th Street, New York, NY 10001

Typeset in Times by Florencetype Ltd, Stoodleigh, Devon
Printed and bound in Great Britain by
Biddles Ltd, Guildford and King's Lynn

British Library Cataloguing in Publication Data
A catalogue record for this book is available from
the British Library

Library of Congress Cataloguing in Publication Data
A catalogue record for this book has been requested

ISBN 0–415–10628–1 (hbk)
ISBN 0–415–10629–X (pbk)

Contents

Illustrations

Preface

Occupational Health: A Practical Guide for Managers is designed to assist managers who recognise the advantages of good occupational health practice within their organisation and wish to develop health-related initiatives.

The book is arranged in four parts. Part I consists of three chapters. The first discusses organisational and legal issues in a way which will facilitate the development of an organisational health policy aimed at the prevention of work-related ill health and the promotion of good health practices among employees. The second chapter traces the history of occupational health and provides information about current practice. Guidance on the development of occupational health policies is contained in Chapter 3.

Part II deals with specific key occupational health issues. Each chapter is devoted to one such issue. As these chapters are designed to stand alone there is inevitably some repetition of background information and, as well as being convenient to the reader, it should serve to emphasise that these issues can never be taken in isolation.

Part III contains chapters dealing with two areas which have substantial occupational health and personnel concerns: the employment of people with disabilities and the employment of women.

Finally, Part IV provides a practical approach to two aspects of health and safety legislation which are having a great impact on all organisations: the Display Screen Equipment Regulations and the Manual Handling of Loads Regulations.

Each chapter contains background information including legal requirements and precedents where appropriate. Ways of implementing relevant occupational health policies are described and in many instances sample policies are provided. The book ends with a bibliography and a list of legal references and useful addresses. At the beginning of each chapter the key points are summarised for easy reference.

Acknowledgements

We would like to thank all our professional colleagues for the encouragement they have given us to produce this book, in particular, Reg Sell, Tony Lackey, Colin Geogeghan, Monty Brill, Barrie Hopson and Dianah Worman, all of whom made invaluable contributions to the format. Our lack of competence on the word processor was compensated for by a team of computer literates including Tharumalingam and Indy Tharmakumar, Judith Laydon, Vivien Stokoe and Jane Parsley. Finally, thanks to our long-suffering spouses, Peter and Stephanie respectively, for putting up with our periods of absence and for acting as readers, and to Ann's daughter for help with the quotations.

Alan would like to dedicate this book to the memory of Vera.

List of abbreviations

ACAS	Advisory, Conciliation and Arbitration Service
ADC	Ability Development Centre
AFOM	Associate Member of the Faculty of Occupational Medicine
AIDS	Acquired Immune Deficiency Syndrome
ASH	Action on Smoking and Health
BAC	British Association for Counselling
BMA	British Medical Association
CBI	Confederation of British Industry
COMA	Committee on Medical Aspects of Food Policy
COSHH	Control of Substances Hazardous to Health
DEA	disability employment adviser
DIH	Diploma of Industrial Health
DSE	display screen equipment
DSS	Department of Social Security
EAP	employee assistance programme
EC	European Community
EDI	electronic data interchange
EMAS	Employment Medical Advisory Service
EOC	Equal Opportunities Commission
ETS	environmental tobacco smoke
FOM	Faculty of Occupational Medicine
HASAWA	Health and Safety at Work etc. Act 1974
HEA	Health Education Authority
HIV	Human Immuno-deficiency Virus
HSC	Health and Safety Commission
HSE	Health and Safety Executive
ILO	International Labour Office
IT	Information Technology
LAYH	Look After Your Heart
LEC	Local Enterprise Company
MASTA	Medical Advisory Service for Travellers Abroad
MAVIS	Master Vision Screener
MFOM	Member of the Faculty of Occupational Medicine
MHSW	Management of Health and Safety at Work

MODU	Major Organisations Development Unit
NALGO	National and Local Government Officers' Association
OHNC	Occupational Health Nursing Certificate
OHND	Occupational Health Nursing Diploma
OSI	Occupational Stress Indicator
PACT	Placing Assessment and Counselling Team
RIDDOR	Reporting of Injuries, Diseases and Dangerous Occurrences Regulations 1985
RCN	Royal College of Nursing
RSI	repetitive strain injury
SOM	Society of Occupational Medicine
SSP	Statutory Sick Pay
TEC	Training and Enterprise Council
TUC	Trades Union Congress
VDU	visual display unit
WRULD	work-related upper limb disorder

Part 1

Chapter 1
Organisational and legal considerations

In three to four years' time, half as many people will be paid twice as much for working three times as hard.

(Charles Handy, *Understanding Organisations*, 1992)

To ensure survival organisations and individuals have experienced, and continue to experience, major changes in the workplace. Technological and cultural advances have led to significant alterations in the nature of work and the demography of the workforce. Leaner, more flexible organisations need new structures and cultures. This chapter details these changes and the driving factors which have led to them. It considers the need for a healthy and effective workforce to meet the challenge of these changes, and discusses organisation structures, cultures, contractual arrangements and the role of the personnel function in relation to these issues.

The changing nature of work

The *Concise Oxford Dictionary* (sixth edition 1976) defines work as 'Expenditure of energy, striving, application of effort or exertion to a purpose'. The eighth edition (1990) changes the definition to 'The application of mental or physical effort to a purpose'. The recognition by the compilers of the increasing importance of mental effort in the world of work in just 15 years reflects the fundamental and rapid changes that have occurred in Britain.

Not only has there been an increase in the non-manufacturing workforce, there has been a significant decrease in the numbers working in the 'making' sector of the economy. Department of Employment statistics show that from a figure of 7 million people employed in the manufacturing industry at the beginning of 1980, by 1992 there were only 4.5 million. Over the same period employment in non-manufacturing had risen from 16 million to 17.2 million (note the increase in unemployment). These figures are not presented in the mistaken view that manual jobs only occur in manufacturing or that manufacturing industry does not employ people whose work is not entirely or mainly mental in its nature. They are presented to support the view that the change from physical to mental work has been rapid and substantial.

Factors driving change

Social and political factors have influenced changes in the types of need being satisfied. For example, in the area of retirement pensions the pressure to make personal provision with less reliance on the state has been a factor in the growth of the financial services sector. The two biggest factors in the change, however, have resulted from technological advances: one in the increased use of machinery to replace manual functions with the parallel increase in machine control; and the other in the availability and sophistication of electronic information processors. These changes have had a fundamental effect on work. It is no longer necessary for employees to travel into an office to carry out their tasks as electronic data interchange (EDI) has facilitated the growth of remote working. Technology has been a major influence in other changes to work. Return on capital investment considerations have influenced working patterns to the extent that shift working of the rotating kind is practised in some 12 per cent of companies in the UK (Blick Time Systems Study 1993). The same study showed that, while 75 per cent of companies had fixed hours, the other 25 per cent had some kind of flexibility.

Technology has also played a major part in removing some of the chemical and physical hazards in the workplace, although other risks have resulted; for example, the increased risk of stress-related illness brought about by the increase in machine rather than human control and the reduction in natural breaks in the working day (or night). Some physical hazards have also been introduced: for example, control equipment which monitors and automatically adjusts many manufacturing processes often contains radioactive isotopes. Operators and maintenance personnel must be badged and monitored to ensure their continued well-being.

The introduction of total quality concepts into British industry has added further impetus to the concepts of continuous improvement, life-long training and competitive edge.

There are counter pressures: one pulling in the direction of increased flexibility with the consequent increase in part-time work and a reduction in demarcation practices, the other pulling in the direction of increased specialisation. The latter has led to an increase in the amount of subcontracting which has occurred in recent years. It is not unusual for an organisation to concentrate on its core activity and contract out activities outside its specialism to other enterprises who are themselves specialists in their function. Examples of this are to be found mostly in functions such as catering, security, transport and warehousing. One of the effects of this is to have 'mixed workforces' – parts of the workforce on any one site being responsible to two or more employers. One of the perceived benefits of such a policy is to pass on to the subcontractor the problem of coping with a reduction in need when economic activity levels drop. This in turn has led to the growth of temporary and part-time employees active in the UK economy. Political factors have amplified these trends by the requirement to seek competitive tenders to market test and to transfer functions from public to private providers.

Competitive pressures have led to 'slimming down'. Cost reduction is not the only factor bringing about this change. Technology not only allows but in many instances requires the operator to have the information, training and authority to carry out the function, thus effectively taking layers out of organisations so that the structures become flatter.

Figures published by the *Employment Gazette* (see *Personnel Today* 1994: 56) show that there has been a tremendous fall in the number of days lost through strikes in the UK. See Table 1.1.

Table 1.1   Number of days lost through strikes

Year	Days/1000 employees
1983–85	587
1986–89	150
1990–92	43

Commenting on these changes, *Personnel Today* (1993b: 48) said: 'Fewer workers' pay was [now] determined by Union agreements and industrial action was less common but dismissals were more common and managers [have] won autonomy.'

The recession, slimming down, subcontracting and other factors which impact on manning levels have resulted in the growth in unemployment leading to less job security. It also has to be noted that unemployment and lack of job security have occurred in social classes, occupation types and geographical locations which traditionally have not experienced such phenomena.

Changes in legislation since 1979 have been a factor in another aspect
of work – trade union membership. In April 1993 *Personnel Today*
(1993a: 48) featured a report by the International Labour Office showing
that trade union membership was declining across most industrialised
economies. The UK has shown the sharpest fall, to 39 per cent from
55 per cent in 1980. An earlier study by the Employment Department
showed that manual closed shops had fallen from a level of 25 per
cent of workplaces in 1980 to 5 per cent in 1990 and that union membership had declined over the same period by 15 per cent (from 75 per
cent to 60 per cent). There has also been a 10 per cent decline in workplaces where unions are recognised by management (65 per cent to 55
per cent).

Demographic considerations

The *Labour Force Survey* shows that the number and gender of people
working changed significantly between 1984 and 1993 (Employment
Department Group 1994). See Table 1.2.

Table 1.2 Population employed: men 16–64 and women 16–59

	1984	1993	%
	Millions		Change
All persons employed	33.125	34.110	+3.0
Men – full time	12.859	12.382	–3.7
Women – full time	5.143	5.871	+14.2
Men – part time	.384	.709	+84.6
Women – part time	3.924	4.593	+17.0

Source: Employment Department Labour Force Survey – spring each year (Crown copyright).

Full-time female employment has increased by 14 per cent, whereas
part-time male employment has increased by 85 per cent, although the
latter increase was from a very low base and reflects a decrease for those
in full-time work of nearly 0.5 million. The figures all reflect a trend
towards the increasing employment of women set against the decreasing
full-time employment of men.

There is a significant reduction in the number of employed persons in the
age group 16–24 (1.2 million) whereas the employed population between
the ages of 25 and 49 has increased by 2.4 million. See Table 1.3.

Table 1.3 Working population by age

Age	1984	1993	%
	Millions		*Change*
16–24	7.949	6.791	–14.6
25–49	17.716	20.129	+13.6
50–64 men 50–59 women }	7.459	7.190	–3.6

Source: Employment Department Labour Force Survey – spring each year (Crown copyright).

The same survey also sought to discover whether health problems were associated with unemployment. There was a large increase overall of 23 per cent in employees reporting health problems. In the relatively small group of part-time male workers, this rose over 200 per cent.

Types of organisation and the organisational structure

Many models have been identified but perhaps one of the most widely used has been the concept of ownership whether public, private, partnership or co-operative. Size, location, process and product are also relevant. While these will impact on work and the surroundings in which it is carried out, it is the effect that these have on the employees and the organisation that is important.

Most organisations have a written company structure, commonly in the form of a family tree. While this is a starting point, other than formal reporting lines and job titles, little information on the organisation and how it functions can be obtained from such charts. Some organisations use the flow diagramming techniques developed by information technology systems analysts to chart the business processes carried out in their concerns (see Figures 1.1 and 1.2). This yields more information and it is now easier to spot weaknesses, duplications and opportunities for improvements than hitherto. (It is also illuminating to note the changes made to the flow diagram when audited by those who actually perform the tasks and know what happens in practice.) A simple example is given at the end of the chapter (see page 11). In many organisations job specifications (or descriptions) are prepared and often used as the basis for recruitment advertisements, job evaluation or grading schemes, induction and training manuals, performance appraisal and other procedures. In recent times, however, the case for laying down the boundaries of someone's job in this manner has been challenged. It is suggested that demarcation and a lack of flexibility are induced in the job-holders, while the cost and effort involved in their preparation and maintenance is not repaid in benefit terms. Tom Peters (1985) wrote: 'There is no greater waste of time than the endless hours devoted to drafting and administering job descriptions.' Alternatives which are finding support include target-setting with mutually agreed performance measures or an agreement as to what has to be achieved expressed in a time-phased way. In other situations work has been organised in a cellular way with the people multiskilled and the team leader operating in the role of trainer and facilitator. The former is heavy on the 'what', while the latter also embraces the 'how'.

Organisational culture

The factors which help to determine structure in turn contribute to an organisation's culture(s). Individual parts of one enterprise will not necessarily have the same culture, as can be illustrated by the cultures found in the production and research parts of most manufacturing companies. It has been suggested that there are four possible cultures: power, role, achievement and support.

Power culture can be described as one where the people in control use resources to satisfy or frustrate the needs of the people in the organisation and thus to control their behaviour. People in power cultures are motivated by rewards and punishments and by the wish to be associated with a strong leader. The power culture is highly dependent upon the leader and, whereas with the right leader it can provide clear direction and quick decision-making, it can also often be accompanied by fear on the part of the subordinates, particularly if they have to impart bad news to the leader. This fear can be detrimental to the health of the subordinates.

The *role* culture substitutes a system of structures and procedures for the naked power of the leader. The struggle for power is moderated by the rule of law. Individuals have defined roles, duties and rewards, while those at each level in the organisation are expected to adhere to their part of the bargain – to perform specific functions for a defined reward. The structure, routine and predictability present in the role culture provide members with security and reduce stress, but can lead to frustration if individual talents and creativity are not fully utilised.

The *achievement* culture has also been called the 'aligned' type of organisation because people are lined up behind a common vision or purpose. In such an organisation systems and structures serve the mission and are changed when the mission requires it. The personal energy of individuals is focused on the mission and they are committed to a shared and common goal. High morale and a 'feel good' factor, springing from being a member of a group and the ability to use one's talents engendered by the achievement culture, can be offset by an intolerance of personal needs where these, family and social life, and quite possibly health, are sacrificed in the interests of work.

The *support* culture is defined as one based on mutual trust between the individual and the organisation. In such an organisation people believe that they are valued as human beings, not just cogs in a machine or contributors to a task.

The support culture, because it is caring and responsive, is nurturing to members and this is good for health, although covert conflict can be present and ambitious people can become frustrated when those making unequal contributions are rewarded equally.

Contractual arrangements

In most employer/employee relationships there is a contract. Indeed, to all intents and purposes employees now have a legal right to a written contract, the elements of which are prescribed by law – wage, hours, holiday entitlement, etc. In the context of occupational health it is the psychological contracts that need to be identified. The organisation has needs but so do the individuals within it. The psychological contract is a set of expectations and results which one contracts to provide to the other. A simple example is that the employer will provide promotion opportunities in return for outstanding performance.

It has been suggested that there are three basic types of contract: coercive, calculative and co-operative (Handy 1976: 40).

As with culture it would be surprising if all the psychological contracts in an organisation were of one type. It is hoped that coercive controls would not be found in too great a number of industrial and commercial enterprises. None the less, it would be less than honest if it were suggested that fear was not still a factor in organisations. Most prevalent will be the calculative contract where the contract is freely entered into and there is a fairly explicit exchange of money for services rendered and control is retained by the management, mainly because they can give desired things to the individual.

Under the co-operative type of contract there is more sharing. The management does not become involved in day-to-day control but seeks to have the employee identify with corporate objectives and, having done this, gives the employee freedom to decide the 'how', the role of management being to provide leadership, training and resources.

While there are literally dozens of specialised symbols used in flow diagrams, most process flow diagrams are built from the following set of basic symbols.

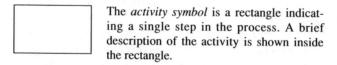 The *activity symbol* is a rectangle indicating a single step in the process. A brief description of the activity is shown inside the rectangle.

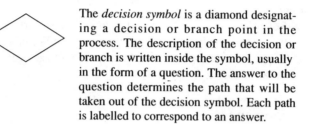 The *decision symbol* is a diamond designating a decision or branch point in the process. The description of the decision or branch is written inside the symbol, usually in the form of a question. The answer to the question determines the path that will be taken out of the decision symbol. Each path is labelled to correspond to an answer.

The *terminal symbol* is a rounded rectangle identifying the beginning or the end of a process. 'Start' or 'End' is shown inside the symbol.

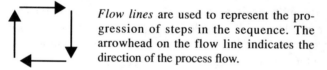 *Flow lines* are used to represent the progression of steps in the sequence. The arrowhead on the flow line indicates the direction of the process flow.

The *document symbol* represents written information pertinent to the process. The title or description of the document is shown inside the symbol.

The *data base symbol* represents electronically stored information pertinent to the process. The title or description of the data base is shown inside the symbol.

The *connector* is a circle used to indicate a continuation of the flow diagram. A letter or number is shown inside the circle. This same letter or number is used in a connector symbol on the continued flow diagram to indicate how the processes are connected.

Figure 1.1 Symbols used in flow diagramming

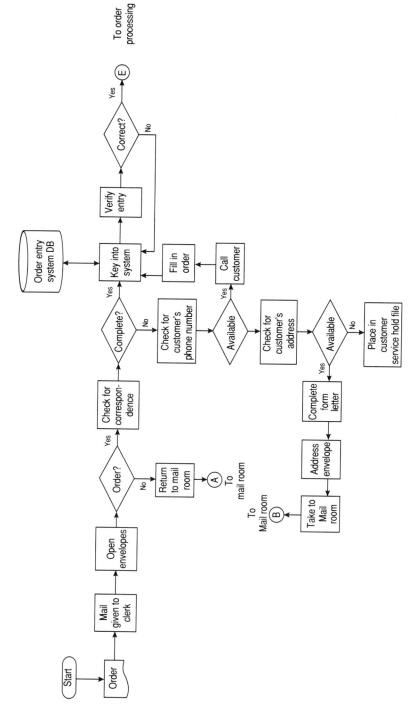

Figure 1.2 A detailed flow diagram: order, receipt and entry
Source: Plsek and Onnias (1989)

The role of the personnel function

Status and scope

Not too long ago – so a story goes – one executive asked another if he was familiar with the three great lies. When he admitted that he knew of two – 'The cheque is in the post' and 'Yes, I'll still love you in the morning' – but could not recall the third, the inquisitor volunteered: 'I'm from Personnel; how can I help you?' This story encapsulates the suspicion with which many personnel departments are viewed. The reasons are many and varied but the fundamental reason is that there is no common understanding of their role, status, authority and boundaries.

Before the Second World War many personnel departments had their origins in welfare sections and their role was seen as concerning themselves with and advising on the well-being of employees. As the needs of the employees, material as well as psychological, were often at variance with the needs of the employer it is not difficult to understand how conflict and misunderstandings occurred.

During the 1950s, the 1960s and the 1970s the role had in many instances taken on a bias towards industrial relations, recruitment, record-keeping and administration, training and development, and health and safety. Even then the authority level was often ill-defined and the role (executive or advisory) confused. With regard to status, personnel staff were often regarded as having a status akin to senior supervisors and convenors or shop stewards. It is now generally recognised that the function is a managerial one and its status is reflected by its level of involvement in the strategic direction of the company, its operational role in the day-to-day management of the company, the level of expertise it brings to its role and the contribution made to the well-being of the people employed in the enterprise and the enterprise as a whole. The breadth of activities embraced within the function differs from company to company. The health and safety specialist, for example, is not universally part of every company's personnel department.

Championing the case for good occupational health practice

Important roles for the function include gaining and communicating senior management commitment to good occupational health practices within the organisation. This involves an understanding of how the organisation will judge and recognise senior management commitment as well as how and by what methods the information flow can be devised and managed. Key elements in the presence of commitment are a consistency of approach, a willingness to provide resources, and a willingness to subordinate the short term in the interests of the long term.

The personnel function should provide an understanding of the status quo and impart knowledge which will allow progress to be made. Progress depends on an ability to overcome resistance, solve problems (and, some would say, circumvent them) and create a desire to make change. There may be a need to convince the organisation, management as well as managed, that the development of a health plan, particularly one embracing mental health, is a legitimate business activity.

As early as 1962, Harry Levinson observed:

> Some among management will scorn their colleagues who express an interest in mental health. These are the men who will say that human relations efforts in industry have failed, that a concern for the health of people is a form of 'softness' not appropriate in industry. If, however, we look closely into situations where human relations practices are alleged to have failed, we see invariably that what passed for human relations was manipulation. The allegation really means that those who sought to manipulate others failed in the effort, and when the psychological confidence game failed, they gave up altogether. Psychological understanding cannot fail. Although there is yet much to be learned and understood, there is already a significant body of psychological and social science knowledge. Management fails when it tries to substitute make-believe for the understanding which can come from this knowledge.
>
> (Levinson 1962)

Other inputs to the knowledge bank can be in the provision of the employment statistics and policies of the organisation – numbers, occupations, age groups, attendance and labour turnover statistics, written and unwritten policies on equal opportunities, pension, sickness absence and related provisions.

Team selection

The function should also assist in the selection of any teams. Ideally, knowledge about team roles and people type should be allied to the knowledge of the individuals under consideration. Useful in this context are the eight team role types identified and described by Belbin (1981):

Company Worker: Conservative, dutiful, predictable. Possesses organizing ability, practical common sense, hard-working, self-discipline. Can lack flexibility and be unresponsive to unproven ideas.

Chairman: Calm, self-confident, controlled. Possesses a capacity for treating and welcoming all potential contributors on their merits and without prejudice. A strong sense of objectives. May be no more than ordinary in terms of intellect or creative ability.

Shaper: Highly strung, outgoing, dynamic. Possesses drive and a readiness to challenge inertia, ineffectiveness, complacency or self-deception. Can be prone to provocation, irritation and impatience.

Plant: Individualistic, serious-minded, unorthodox. Possesses genius, imagination, intellect, knowledge. Can be up in the clouds, inclined to disregard practical details or protocol.

Resource Investigator: Extroverted, enthusiastic, curious, communicative. Possesses a capacity for contacting people and exploring anything new. An ability to respond to challenge. Can be liable to lose interest once the initial fascination has passed.

Monitor Evaluator: Sober, unemotional, prudent. Possesses judgement, discretion, hard-headedness. Can lack inspiration or the ability to motivate others.

Team Worker: Socially orientated, rather mild, sensitive. Possesses an ability to respond to people, and to situations, and to promote team spirit. Can be indecisive at moments of crisis.

Complete Finisher: Painstaking, orderly, conscientious, anxious. Possesses a capacity for follow-through. May be a perfectionist and have a tendency to worry about small things. May have a reluctance to 'let go'.

(Belbin 1981: 78)

Recruitment and selection

Where there is a specialist personnel function it is that function which would normally provide a recruitment service embracing short-listing, candidate testing (where appropriate) and the expression of opinions on individual candidates. It is the line manager to whom the recruit will report who carries out the selection. The executive decision with regard to acceptance or rejection of candidates is, therefore, vested in the 'user'.

Recruitment is concerned with identifying and tapping the sources from where it is most likely that the successful candidate will come. With some appointments a source of candidates may well be from within the organisation or from speculative (or previously unsuccessful) applicants stored within the records of the personnel department. Other appointments may best be sourced by advertising or from specialist agencies. The acid test of selection is the job performance level achieved by the successful candidate, although it could be argued that the rejected candidate(s) may have done even better. The importance of person/job fit in reducing stress cannot be over-emphasised. Other tests of effectiveness and efficiency are the satisfaction level of the selector and of the candidates (both successful and unsuccessful), compliance with policy and equal opportunities, clarity of the selection criteria, speed and cost.

The personnel function should consider how well the recruitment and selection procedures stand up to these tests by looking at failure rates, subsequent career moves of individuals and any relationship to the recruitment source. The procedures should facilitate decision-taking and be economic both in terms of executive time and money.

It is strongly recommended that the recruitment and selection process is mapped out using flow diagramming, preferably time phased, and that the audit working party (see Chapter 3) also has access to information gained from exit interviews, which any professional personnel specialist will ensure are carried out when people leave the organisation.

Training and development

The training and development aspect of personnel work is often absent in occupational health considerations, in contrast to provision for training in safety. The role that occupational health professionals can play in training and the need for training in occupational health issues are referred to in Chapters 3, 11 and 12.

Health and safety legislation

Since the introduction of the Health and Safety at Work etc. Act (HASAWA) in 1974 there has been increasing legislation on health and safety.

Landmark regulations stemming from this which have had an impact on many industries are the Control of Substances Hazardous to Health (COSHH) Regulations 1988, the Noise at Work Regulations 1990 and most recently the 'six pack' of regulations emanating from the European Community Framework Directive:

- Management of Health and Safety at Work Regulations
- Work Equipment Regulations
- Manual Handling of Loads Regulations

- Workplace Health, Safety and Welfare Regulations
- Personal Protective Equipment at Work Regulations
- Display Screen Equipment Work Regulations.

The COSHH Regulations have been perhaps the most significant in the development of occupational health and safety measures. A CBI survey in 1993, which showed that 50 per cent of private sector employers had only introduced such measures over the last five years, seems to reflect this trend. To many employers the implementation of safety regulations has been seen as, and indeed is, a burden on both staff and financial resources. Although alterations in work methods, practices, equipment and environment may prove costly, probably the greatest problem has been assessment. These regulations all require an assessment to be made of the risk by a competent person. In organisations which do not have trained safety or occupational health staff it has been difficult to set up suitable assessment programmes. Alternatives have been to:

- appoint a trained safety officer;
- appoint a trained occupational health practitioner;
- employ health and safety practitioners on a consultancy basis;
- train designated staff.

There is a large gap between legislation and its enforcement, and many organisations continue to break the law with regard to health and safety. The Health and Safety Executive's approach has been to persuade employers to achieve what is reasonably practicable. This pragmatic *modus operandi* is resulting in a gradually increasing degree of conformity. At a time when most organisations have major financial constraints, health and safety initiatives do need to be seen as realistic.

The concept of risk seems difficult for many managers to grasp. It is simply the likelihood of a known hazard causing damage to health. Another dimension to this assessment is the numbers of staff, contractors and the public who may be at risk. The Management of Health and Safety at Work (MHSW) Regulations have closed any gap that might have been supposed to exist in the list of hazards which have to be assessed. Although the HASAWA did require protection of staff, assessment of risk was tied to specific regulations. The MHSW Regulations require the assessment of risk for *all* workplace hazards. Of course, those organisations which are complying with previous industry and hazard specific regulations will find that they have already covered most areas. Those which are not will have to set in train the full assessment programme. It is certainly not possible to address this issue without the use of trained health and safety experts.

One of the main features of the HASAWA was to be the involvement of staff representatives in health and safety committees and in safety inspections. The trade union movement provided a successful training programme for its representatives. Unfortunately, as more workplaces become less unionised and there are fewer trade union members trained, staff representation in joint committees is becoming increasingly difficult to sustain. Organisations will probably need to address the issue to determine ways of fully involving the workforce in compliance with the legislation.

There is no doubt that individuals have a responsibility for their own health and safety, and must follow safety procedures, but equally managers must ensure that these are in place and are practical.

Chapter 2
An introduction to occupational health

He shortens his life and he hastens his death
Tally hi-o, the grinder
Will drink steel dust in every breath . . .
Won't use a fan as he turns his wheel
Won't wash his hands ere he eats his meal
But dies as he lives as hard as steel
Where rests the heavier weight of shame?
On the famine-price contractor's head
Or the workman's under-taught and fed
Who grinds his own bones and his child's for bread?

(Anon.)

The practice of occupational health today has evolved from the work of Thackrah in the early nineteenth century. Government, employer and later employee initiatives drove forward the concept of employee health and safety, particularly in relation to physical health. Following the Second World War two new developments shifted the focus, and shaped the practice of occupational health practice today: the recognition that mental health was of significance in the workplace; and the use of health promotion techniques to produce a more effective workforce. This chapter describes briefly the development of occupational health in the UK which has led to the present rather patchy provision. Two surveys of occupational health practice commissioned by the Health and Safety Executive (HSE) are discussed in detail. The chapter ends with a summary of available occupational health activities and advice on how to develop an occupational health service.

History of occupational health

The earliest recorded example of medicine within the industrial setting was around 1600 when a medical attendant was provided for miners in Tintern. Bernardino Ramazzini, a Professor of Medicine at the University of Modena, is considered to be the father of occupational medicine. During inspections relating to public health issues he became concerned with the well-being of workers and published *De Morbis Artificum Diatriba* in 1700, in which he described diseases related to occupations as diverse as mining and singing. However, his special contribution to general medical practice was the imperative that when taking a patient's history a doctor should always ask the question, 'What is your occupation?'

It was an Englishman, Charles Turner Thackrah, however, who established the idea of occupational medicine as a speciality. A physician based in Leeds, in 1830 he published *The Effects of the Principal Arts, Trades and Professions and of Civic States and Habits of Living on Health and Longevity, with Suggestions for the Removal of Many of the Agents which Produce Disease and Shorten the Duration of Life.* Among many other conditions, he identified lung disease in miners and grinders of metals, postural deformities in young working children and lead poisoning in glaze dippers. It is difficult to know how great his influence was on the legislators of the day but his publication was followed shortly by the first official introduction of a medical opinion into the workplace. The Factory Act of 1833 required that a surgeon should certify that a child was of the strength and appearance of a nine year old. Owing to misuse of this system it was enhanced in 1844 when more specialised certifying surgeons were appointed.

The development of occupational medicine has been on three fronts: employer-controlled schemes, worker-controlled schemes and government initiatives through the work of the Factory Inspectorate.

Employer initiatives

The Post Office Medical Service was formed in 1855 and between then and the end of the nineteenth century other employers, notably in the railways and the Quaker confectionery industry, established medical services. Management-funded employee welfare schemes developed rapidly between the First and Second World Wars, stimulated by organisations such as the Industrial Health Research Board and the Institute of Industrial Psychology. The concept that safety, health and welfare were an important aspect of management began to take root with the developing ideas of scientific management (Taylor 1911). A chilling example of health and safety management was seen in Nazi Germany where there was increased recognition and prescription of industrial disease, but this ran parallel with increased restriction of workers' rights. New and powerful full-time industrial medical officers were appointed (often highly placed party officials), not only to tighten up on supervision of safety controls but also to discourage workers from taking sickness absence.

Employee initiatives

Parallel to the employers' initiatives there were a number of worker-controlled developments. The Amalgamated Society of Engineers was formed in 1851 and the trade union movement continued to grow in power until the Thatcher governments of the 1980s. Unfortunately, too often union negotiations were concerned with reward rather than safety (Kinnersley 1973). In the late nineteenth century there was a proliferation of friendly societies which, in return for a regular subscription, provided payment for medical treatment. In certain areas this was much more advanced; for example, miners in South Wales developed and controlled an extensive medical service. The interest of the Trades Union Congress in the general health of employees culminated in the funding and use of the Manor Hospital for trade union members. The establishment of the TUC Centenary Institute of Occupational Medicine in London in the 1970s was a major step forward in the recognition of the genuine occupational health needs of the workforce.

Government initiatives

The 1879 Report of the Chief Inspector of Factories for the first time recognised 'occupations injurious to health' and in 1878 Thomas Legge was appointed as the first Medical Inspector. The Factory Inspectorate was, and continues to be, a major influence in achieving healthy working conditions. It has always been hampered by lack of personnel with an inevitable gap between legislation and enforcement. From an early stage, however, it has seen its role as attitude-forming and educational rather than punitive. It is perhaps unfortunate that the ministry concerned with labour (the nomenclature changes) is the one that has also been given the task of responsibility for the health of the workforce. This has proved to be a barrier to the development of a comprehensive national occupational health service. Practitioners in occupational medicine continue to argue for such a service under the umbrella of the Department of Health and in parallel with the National Health Service. With the establishment of the Health and Safety Commission in 1974 and the appointment of a substantial number of employment medical advisers, it was hoped that there would be better co-ordination. This has not been the case and occupational health provision for workers still remains patchy.

Occupational health today

Who has it?

The Employment Medical Advisory Service (EMAS) undertook the first national survey of occupational health services in 1977: *Occupational Health Services – The Way Ahead*. With the proliferation of new legislation, changes in the distribution of the workforce, the introduction of new procedures and increased awareness of health and environmental issues, a further survey, *Occupational Health Provision at Work*, was commissioned in 1993. The results of this are discussed below.

It is clear that size is the key determinant of whether an establishment takes any occupational health measures and the range and scope of these measures. See Table 2.1.

Table 2.1 Occupational health measures in the private sector

Number of employees	%
<10	58
11–24	82
25–49	89
50–199	93
200+	100

While size is the principal factor, the type of industry is also important, 78 per cent of the production sector and 60 per cent of the service sector having some occupational health initiatives. Around 89 per cent of the working population have access to some occupational health measures.

Seventy-four per cent of the employers in the private sector, and nearly all public sector employers, acknowledge responsibility for employees' health and safety. In both sectors there was a surprisingly high degree of confidence in the adequacy of the occupational health provision (around 95 per cent in the private sector and 82 per cent in the public). Employees, on the other hand, were not as confident, 27 per cent believing that the employer could do more.

In-depth questioning was far less encouraging in terms of knowledge of health and safety issues. Fewer than half of those questioned in the private sector were able to suggest spontaneously any actual hazards to which their workforce was exposed. This rose to 75 per cent in the public

Table 2.2 Awareness of hazards in the workplace in the private sector

Number of employees	%
<10	38
11–24	48
25–49	60
50–199	66
200+	76
Overall	42
Industry type	
Production	57
Service	37

sector. It is difficult to see how the acknowledged responsibility for health and safety can be addressed if the knowledge of hazard is so poor. See Table 2.2.

Employees are certainly aware that work can damage their health. Thirteen per cent believed that they were suffering from significant work-related problems, back pain and stress being the most common.

Reasons for occupational health provision

Approximately 35 per cent of private sector employers have no occupational health provision.

These employers showed no particular interest in introducing any occupational health measures. When prompted, several reasons were given which would lead them to do so. See Table 2.3.

Table 2.3 Reasons why occupational health measures might be introduced

Reason	%
Health problems at work	38
Visit from HSE etc.	9
Employee pressure	8
Legislation	6
Required by manufacturer/supplier	3
Required by customer	1
Don't know	44

The rather weak motivation induced by the Factory Inspectorate perhaps reflects the gap between legislation and enforcement already considered in this chapter. However, a similar survey by the Confederation of British Industries (CBI) (*Working for Your Health*) also in 1993 gave a slightly different picture. See Table 2.4.

Table 2.4 Motivation for concern about employees' health

Motivator	Average ranking
Potential health and safety risks	2.2
Health and safety legislation	2.9
Potential absence costs	3.2
Improving employee morale	3.6
Reducing employee turnover	4.8
Promotional companies	5.6
Insurance premiums	5.6

Employers were given a more structured questionnaire in this survey. They were asked which pressures motivated their concern about the health of their employees and were given a list to rank from 1–7 with 1 being of greatest importance. The results showed a greater emphasis on the effects of employee ill health and health education. Employers also indicated that they were influenced by legislatory requirements. Insurance considerations are of increasing significance.

A different picture emerges when employers were asked why they had introduced occupational health measures. See Table 2.5.

Table 2.5 Reasons for introducing occupational health measures

Motivator	%
General review of health and safety	18
Legislation	11
Visit from HSE etc.	5
HSE literature	5
Health problems at work	5
Employee pressure	4
Advice from trade association	3
Request by customer	2
Request by manufacturer/supplier	1

In this case health and safety factors were identified as major motivators. It appears from this that those employers who already have occupational health measures are more aware of health and safety requirements. This may be a retrospective reading of why they took the initiative.

Type of occupational health provision

The type of occupational health provision varied enormously but COSHH Regulations-related risk assessments, hazard data sheets, safety notices and the provision of protective equipment were the most common measures in both public and private sectors.

The percentage of private sector companies providing formal safety training was low, averaging around 15 per cent.

Use of health professionals

When one looks at the quality of occupational health provision the situation is far from encouraging.

In the private sector less than 8 per cent have input from any sort of health professional. The use of health professionals is higher in the public sector, over 50 per cent of the workforce having access to trained staff. Comparison with earlier figures suggests that there is an increase in the use of health professionals and the survey showed that over half of the private sector using health professionals have only done so during the last five years. There is an indication, however, that fewer than half of the professionals used are trained in occupational health. This seems a somewhat dismal situation, which will be considered later in ascertaining why occupational health has failed to make a more general impact on employers and employees. Although the Association of Industrial Medical Officers was formed in 1935 (now renamed the Society of Occupational Medicine) and the Faculty of Occupational Medicine was launched in 1977, occupational medicine has still to establish itself as a speciality in the eyes of the general populace.

Professional advice on aspects of occupational health is available from a number of different specialities:

- occupational health doctor;
- occupational health nurse;
- occupational hygienist;
- occupational psychologist;
- ergonomist.

However, the survey showed that coverage of the workforce by these specialists was low, particularly in the private sector. The situation is in reality much worse since the figures do not reflect the considerable number of doctors and nurses still working in occupational health who have no formal qualification in the subject. Fifty-three per cent of those employed in the public sector were described as having occupational health qualifications as compared with only 19 per cent of those in the private sector. Just 20 per cent of private sector employees have access to even a part-time doctor and surprisingly only 14 per cent have access to a part-time nurse. The public sector figures are more encouraging (72 per cent and 86 per cent respectively).

The CBI survey also sought to determine from whom organisations obtained health and safety advice. Of those surveyed 40 per cent relied on a safety officer, 34 per cent on the in-house doctor, 25 per cent on EMAS and 22 per cent on an occupational health consultancy.

Role of occupational health

Since the establishment of the Association of Industrial Medical Officers in 1935, the role of occupational medicine has been constantly evolving, as reflected in the change of name to the Society of Occupational Medicine. Discussions are currently under way with a view to adding 'Environment' to the title. It is not surprising that occupational medicine has changed since it rightly reflects the needs of the workforce. As we have seen in Chapter 1, the workforce has changed substantially in the nature of work undertaken and its organisation. It is also affected by increasing health and safety legislation. Added to this are the expectations of the workforce of a safe place of work, where there are opportunities for personal development. The most recent impetus for change has come from the Department of Health (1993) – *The Health of the Nation* initiative, specifically targeting health at work.

Perhaps because of the continuing use of staff unqualified in occupational health in occupational health departments, the activities of many units have a strong bias towards the clinical aspects of occupational health; that is, those directly relating to the health of the individual employee, such as pre-employment screening and treatment and, more recently, health education. See Table 2.6.

In fact, occupational health practitioners can offer much more to an organisation:

- *Assessing fitness for work*
     Recruitment;
     Rehabilitation;

Table 2.6 Occupational health measures

	% Public sector	% Private sector
Medicals	89	42
Regular health checks	83	39
Advising on health and safety measures	82	43
Attending health and safety meetings	82	27
Identifying other areas which may cause problems	72	40
Monitoring health and safety procedures	64	66
Monitoring sickness absence records	54	22
Implementing health and safety procedures	51	25
Management of health and safety	47	26
Carrying out COSHH Regulations	37	28
Treating ill health and accidents	18	4

Resettlement;
Retirement.
- *Assisting in the implementation of health and safety legislation*
   Developing policies;
   Advising management on responsibilities;
   Developing and providing training programmes;
   Assisting in risk assessments;
   Ergonomic assessments;
   Undertaking statutory examinations;
   Medical surveillance;
   Biological monitoring;
   Screening;
   Accident review;
   Assisting in the epidemiological management of sickness studies
      absence;
   First aid training.
- *Health promotion*
   Individual consultations;
   Policies;
   Campaigns;
   Workshops;
   Classes;
   Literature.

- *Stress management*
    Counselling workshops;
    Support groups.
- *Treatment*
    Emergency (with first aiders);
    Routine;
    Using other health professionals;
    Therapeutic massage;
    Physiotherapy;
    Osteopathy etc.
- *Research*

Occupational health activities

Recruitment

It is now generally recognised that pre-employment medical examinations are not cost effective. A recent study within the National Health Service showed that of 9139 pre-employment health assessments only 2 per cent were found unfit by whatever means of selection and a further 1.3 per cent were considered fit for restricted duties. The rejection rate using a health questionnaire was the same as the rejection rate following a pre-employment health check by a nurse. Of the 688 who were selected from the information on the health questionnaire as requiring a nurse interview, only 25 were subsequently rejected. Of the 390 referred for medical assessment only 52 (6.2 per cent) were found unfit.

Of course, certain occupations have statutory requirements to ensure fitness and there are others where a reasonable employer would also wish to ensure fitness for safety reasons. The pre-employment health check may be used for a number of other purposes such as establishing baseline data, introducing the occupational health service and discussing safety issues. It is important to be clear about the true purpose of recruitment health encounters. The stethoscope in this situation rarely reveals anything that cannot be determined by past history of illness and of attendance. Indeed, the likelihood of regular attendance is most accurately predicted by work attendance records in the previous two years.

The Clothier report (1994) on Beverly Allitt has raised issues about past history in applicants for nurse training posts and the Royal College of Nursing is currently considering health standards for nurse entrants. From the occupational health point of view, it is difficult to see how personality tests and detailed medical history can act other than as an exclusion rather than a selection process. There is no evidence that any psychological test would identify the dangerous individual at the recruitment stage. Neither is there any evidence that individuals with earlier psychiatric problems make less satisfactory nurses.

Assisting in the management of sickness absence

Occupational health physicians should not be expected to control sickness absence but they are able to assist managers in doing so. The physician should look at the epidemiology of sickness absence, providing statistics on levels of absence in different work groups. This may help to pinpoint not only hazards but also management weaknesses and low morale. It is not always possible to give a realistic prognosis for an individual's attendance at work, but some indication of what can be expected in relation to a particular diagnosis is possible and may assist managers in formulating reasonable attendance standards. With the new arrangements for Statutory Sick Pay it has become increasingly important for managers to deal with sickness absence in a cost-effective way. There is a full discussion of this in Chapter 6.

Rehabilitation and resettlement

Occupational health professionals have a particularly important role to play in this function. They are the only people with knowledge and understanding of the medical condition *and* the requirements of the job. General practitioners are at the mercy of their patients whose description of their jobs may directly relate to their eagerness or reluctance to return to work. Unfortunately, most organisations now seem unable to accommodate the walking wounded for any length of time, making rehabilitation jobs or variations of jobs difficult to organise. Similarly, no one wants to take on from another department someone who appears to have a long-term health problem. This all leads to the unnecessary loss of trained staff and longer-term sickness absence. Occupational health practitioners continually strive to encourage management to be more flexible in this respect.

Assisting in the implementation of health and safety legislation

Where an organisation has a qualified safety officer it would be normal for him to work closely with the occupational health professional, providing management with guidance on the implementation of policies designed to meet the requirements of health and safety legislation. The roles of both professionals are complementary. Where there is no professional safety officer, the occupational health incumbent will provide management with advice, help to develop policies and training programmes, assist in assessments, provide training and undertake any necessary medical surveillance or biological monitoring. For small employers this combination of roles may be the most economical approach.

Accident review
It is normal for accident reports to be copied to the occupational health department. Under the Reporting of Injuries, Diseases and Dangerous Occurrences Regulations 1985 (RIDDOR) the occupational health physician is required to report any industrial disease. The occupational health department expects to follow up any accidents to see that appropriate action has been taken and advise on this. It often falls to the department to provide accident statistics and suggest target areas for safety campaigns.

Health promotion
The advantages and disadvantages of health promotion in the workplace are explained in Chapter 8. Clearly, the working population is a particularly important target in areas mentioned in *The Health of the Nation* document, namely coronary heart disease, cancers and stress. To some extent it is a captive audience and employees should be encouraged, for example, by being given time off to attend workshops and other training programmes, to assume responsibility for their own health. There has been a recent tendency to equate occupational health with health promotion. Although an important aspect of the role, it is perhaps the least obviously cost effective.

Stress management
For a number of reasons stress is now seen as a major problem in most work settings. In the 1993 HSE survey *Occupational Health Provision at Work*, it was rated by employees on a level with back trouble as a cause for concern. A comprehensive approach to the mental health of the workforce is described in Chapter 4.

Treatment
Occupational health is not a substitute for general practitioner delivery of primary care. Fortunately, this is now recognised by both general and occupational health practitioners so there is no misunderstanding between the two professional bodies. Of course, it would be unreasonable for the occupational health professional not to offer advice to those injured or becoming ill at work. In some instances timely intervention may prevent sickness absence and, at the very least, provide individuals with some relief until they can see their general practitioner. Where there is a regular occupational health presence, the general practitioner may request help

with routine treatments, making this easier for the patient and reducing the need to take time off.

Increasingly, other health professionals are being introduced into the workplace to enable employees to obtain treatments conveniently and often at reduced rates. Osteopaths, therapeutic masseurs, Alexander technique practitioners and many others are now part of the occupational health portfolio.

First aid

The First Aid at Work Regulations 1981 clearly specify requirements for first aid provision in the workplace. Organisations are required to assess the degree of risk in their operations and provide adequate first aid cover. HSE guidelines suggest that the presence of an occupational health department does not remove the total requirement to provide first aiders. The location and the hours of work of occupational health staff will almost certainly mean that they cannot provide adequate cover.

However, occupational health can assist in developing and maintaining the first aid provision by:

- assisting in the selection of first aiders;
- organising and running primary/refresher courses;
- organising regular up-date sessions;
- developing accident reporting and procedures;
- developing first aid procedures in relation to special hazards.

Research

Although occupational disease (i.e. specific disease resulting from exposure to a hazard in the workplace) is becoming less frequent, there is still a need for vigilance where chemicals and new technology are in use. Occupational health has a major role in detecting, through surveillance, attendance records and other forms of recording, the incidence of work-related disease. The most recent epidemic of work-related upper limb disorder has been highlighted by occupational health practitioners and they are spearheading its control.

Setting up an occupational health service

With increasing health and safety legislation, new Statutory Sick Pay arrangements, tighter budgets and the need to conserve human resources, organisations are reassessing their occupational health provision. The use of trained occupational health staff helps to focus the needs of the organisation and provides a more cost-effective service. Except in large organisations there is a trend towards the use of external occupational health facilities and the number of such facilities is increasing. Many large organisation-based services are tendering to be providers to smaller organisations. This may be a more effective use of the limited number of trained occupational health personnel and should limit the unsupervised use of untrained personnel. In assessing the occupational health needs of an organisation it may be appropriate to seek expert advice.

Guidance on where this advice can be obtained is available from the Employment Medical Advisory Service, the Society of Occupational Medicine and the Occupational Health Section of the Royal College of Nursing. There is no direct legal requirement for an employer to employ doctors and nurses. However, where there are statutory requirements, such as the provision of medical or biological surveillance, it may be sensible to employ suitably trained practitioners.

Appropriate qualifications for doctors are Associate Membership, Membership or Fellowship of the Faculty of Occupational Medicine (AFOM, MFOM, FFOM), the Diploma of Industrial Health (DIH) or an MSc in Occupational Medicine. In 1994 the Faculty introduced a diploma for doctors who wish to work part time but not specialise in occupational medicine (DOM). For nurses, the most relevant qualifications are the Diploma or Certificate in Occupational Health Nursing (OHND, OHNC).

Chapter 3 discusses how to audit an organisation to determine occupational health requirements. Management commitment and the availability of resources will further determine the appropriate level of intervention.

Confidentiality

Occupational health requirements vary widely from industry to industry but the basic concepts and ethics remain the same. It should be remembered that occupational health physicians and nurses are bound by strict rules of confidentiality in relation to individual members of staff. 'The status of an occupational physician in an organisation must be that of impartial professional adviser, concerned primarily with safeguarding and improving the health of employed persons' (Faculty of Occupational Medicine).

Occupational health staff are restricted from discussing the clinical history of the patient with a manager unless specifically requested to do so by the patient. For the same reason, medical records are the property of the occupational health department and no other staff may have access to them. Reported information on the patient should not contain details of their medical condition but only information about how that condition affects their ability to work. Managers sometimes find it difficult to accept these principles of medical confidentiality and may take time to understand and feel secure about the occupational health role. Occupational health staff will consider any situation from both the organisation's and the patient's point of view and should not put either at risk.

Chapter 3
The organisational health plan

The test will be less the effectiveness of our material investment than the effectiveness of our investment in man.

(J. K. Galbraith, *The Affluent Society*, Chapter 25: iv)

The good employer now recognises that promotion and protection of employee health is not only mandatory but makes good business sense. But too often employers embark on small and poorly researched occupational health initiatives. In this chapter the development of an appropriate occupational health plan for an organisation or department is described. A checklist is provided to facilitate the assessment of the organisation with regard to occupational health needs. This includes consideration of the culture and structure of the organisation, the nature of the work, selection of personnel, sickness absence and other personnel procedures, training, and the management of health and safety. The components of an occupational health plan are described in five sections: person/job fit, organisational style, health-related policies, staff support systems and health promotion. A model policy is given at the end of the chapter.

Introduction

During the course of the twentieth century the requirements of a good employer have changed radically both on the psychological and physical fronts.

F. W. Taylor's (1911) idea of man as an essentially idle, passive being who can only be motivated to work by financial reward has long been rejected by proponents of management theory. As early as the 1930s man's social motivation to work had been clearly demonstrated. To manage successfully it was seen that man's social needs must also be addressed (Herzberg 1966; Porter and Lawler 1968).

The manager could no longer just be a controller and reward dispenser in a strictly hierarchical organisation. He had also to facilitate a socio-technical system with employee-oriented leadership and different groupings within the management system. Maslow (1970) introduced an additional individual need for some sort of autonomy for the worker.

A parallel development which has affected the physical state of the workplace is that of health and safety legislation. The Health and Safety at Work etc. Act 1974 required employers to provide a safe place of work. Since then there has been continual development in safety requirements, most recently stemming from European Community (EC) health and safety legislation. Employment law has set basic standards for employee welfare which are also being influenced by EC legislation.

The promotion and protection of employees' health is no longer seen as philanthropic. Not only are there mandatory reasons for being a responsible employer but good business reasons to ensure an effective and motivated workforce.

Accepted characteristics of good employment are:

- *Physical*
    A safe workplace;
    A pleasant as possible work environment.
- *Organisational*
    Reasonable hours of work;
    Psychologically acceptable shift patterns;
    Rest periods and suitable facilities;
    Sick pay;
    Resettlement and rehabilitation facilities;
    Opportunities for early retirement;
    Recreational facilities;
    Fair reward and recognition.
- *Psychological*
    Opportunities for personal growth;
    Opportunities to improve skills;
    Opportunities to work autonomously;
    Opportunities for growth outside the workplace;
    Fair management;
    A good appraisal system;
    A system for problem-solving.
- *Social*
    Childcare facilities or support;
    Flexible hours;
    Imaginative work/home arrangements.

It is surprising that, in view of the general acceptance of the ideas of the human relations school of management, plans for employee health are not a common part of any business plan.

Developing the health plan

Developing a plan for individual and organisational health provides an opportunity to assess the needs of the organisation in a structured way and put in place policies and procedures relevant to its activities and culture. An organisational commitment to health has to encompass not only the stated intention to enhance employee health and well-being, but also the provision of resources to carry this forward. Any health plan needs to include initiatives for both individual and organisational health.

The development of the plan should be in four stages:

- Assessment;
- The draft health plan;
- Implementation;
- Audit.

To ensure ownership at an early stage, a small working party should be set up to include a committed senior operations manager, union representative, personnel manager, occupational health and safety adviser and, if not one of the above, the individual who will carry out the assessment. It cannot be emphasised too strongly how important it is to have senior management involvement at this stage.

Assessment

In order to develop a comprehensive plan for employee health, it is necessary first to look at the existing organisational structure and culture. The right person to do such an assessment is likely to be trained in the broad area of human resource management and may be a personnel manager, occupational medicine practitioner or occupational psychologist. Where the expertise is not available in-house, it can be obtained from outside agencies. Areas covered should include:

- the organisational structure and culture;
- the nature of the work;
- the role of the personnel department;
- personnel selection procedures;
- sickness absence procedures;
- health-related personnel policies;

- staff training and development;
- staff representation;
- management of health and safety.

A sample checklist can be found at the end of this chapter (see pages 50–1).

The organisation structure and culture

The deep-set beliefs about the way work should be organised, authority exercised, and people recognised, rewarded and controlled make up the culture of an organisation and determine what sort of people are employed and what their career aspirations are likely to be. It is important to identify the culture of the organisation, which may be power-based, role-based, support-based or achievement-based. The structure of the organisation will depend on this culture and reflects how the organisation is managed and communication is achieved.

It is important to identify not only the current state of the organisation but also any imminent or ongoing changes. In general, contracts of employment are becoming shorter and there are few employees who are in a job for life. This may produce particular stresses where terms of employment have changed significantly. Total quality concepts emphasise the need for continuous improvement and life-long learning.

Few companies now provide the sort of paternalism that was seen in companies such as Unilever in the 1920s and the 1930s when you could be born in a Port Sunlight hospital, go to a Port Sunlight school, work for the company for 40 years and be buried by them. However, there is still a world of difference between the in-and-out quick profit world of the money-broking companies and the traditional philanthropy of organisations such as the John Lewis Partnership. In assessing the organisation's needs the underlying structure and culture will be the key to what is relevant and practical. Chapter 1 gives a detailed account of organisational structures and cultures.

The nature of the work

In deciding on an appropriate health plan probably the two most important aspects of an organisation are its culture and the nature of the work. There are likely to be significant differences in the requirements, say, of a construction company and a bank.

The role of the personnel department

This is likely to reflect the prevailing mores and may be restricted to ensuring that the organisation conforms to employment law and has a satisfactory industrial relations structure. In other organisations there may be a strong pastoral role, where personnel officers are trained as counsellors and are seen as employee representatives. There may be a degree of confusion about the role of personnel. In most cases the emphasis is on a specialist management role (see Chapter 1).

Personnel selection procedures

These are often a matter of custom and practice, and their efficacy may not have been assessed. But an assessment of their efficacy should be part of the development of the health plan. It will include such issues as the appropriateness of the use of practical and psychological tests. Should they be used? Are they valid? What part should they play in selection? Are assessment centres used?

Sickness absence procedures

Sickness absence entitlement for different grades and the effective management of sickness absence are both issues crucial to employee well-being.

The organisation's current procedures and policies should be scrutinised for efficiency and fairness. This will also include procedures for early retirement on the grounds of ill health.

Health-related personnel policies

The assessment should include an examination of existing health-related policies, such as control of alcohol misuse and smoking control.

Staff training and development

Assessment for the health plan should include three areas of training relevant to employee health:

- training to do the job;
- management training;
- general life skills training.

Staff representation

New employment law has lessened the powerful unionisation of many industries and staff representatives may have little negotiating power within the organisation. It is likely that the development of a health plan will be welcomed by staff representatives and they should form part of any working party to develop the plan.

Management of health and safety

The organisation's commitment to employee health may be quickly assessed by looking at health and safety structures already in place. This is not a question of numbers or of accommodation but of what is actually being done, and how relevant it is to the real, rather than the perceived, needs of the staff.

The draft health plan

The assessment period may take some time. This is worthwhile, however, as initial ideas can be tested out at later meetings and in group discussions. It is important not to form opinions too soon but to listen to 'the music behind the words'.

Procedures already in place which address the health needs of staff will have been identified, as will areas where there is a need for further development.

The plan is divided into five parts:

- Improving the person/job fit;
- Developing a healthy organisational structure and culture;
- Developing health-related policies;
- Developing staff support systems;
- Health promotion.

Improving the person/job fit

Personnel selection procedures

It goes without saying that there are 'horses for courses' and that the right person in the right job is a recipe for both individual and organisational well-being. However, it may be difficult to ensure the person/job fit.

Two kinds of fit between the individual and the environment must be considered: the extent to which the individual's skills, abilities and attitudes match the demands and requirements of the job and the style of the company, and the extent to which the job and its psychological environment meet the individual's needs. To ensure a person/job fit, it is important to take into account not only the objective view of the person and the environment but also the individual's view of his abilities and the environment.

For most jobs the profile of the perfect employee has not been defined. If psychological tests are to be used a profile must be established, so that the test can effectively measure elements of this profile.

The personnel specification normally includes seven points recommended by the National Institute of Industrial Psychology:

- physical characteristics;
- attainments;
- general intelligence;
- special aptitudes;
- interests;
- disposition;
- circumstances.

Psychological testing is used to assess motivation, attitudes, intelligence and abilities, and personality. There is little information on the effectiveness of such tests which are widely used, particularly in management appointments. When considering the selection procedures, the working party should identify the areas where psychological testing may be appropriate and assess the validity of any proposed tests.

Following cases such as that of Beverly Allitt there have been demands for psychological screening. These demands usually fall short of identifying the personality factor which might be a valid measurement.

Practical tests may be considered where an aptitude for certain functions, such as fine manual handling or colour discrimination, is important. Care should be exercised when interpreting the test results. For example, train drivers or electricians should not necessarily be excluded from such work because of imperfect colour vision, demonstrated by the Ishihara colour charts, if they are able to complete a practical test satisfactorily.

Where there are definite physical fitness requirements for posts, a proper assessment of the requirements should be made and tests put in place which assess these.

The current interviewing procedures should also be examined to ensure that the length of interview is correct and the appropriate people are conducting it, and that those involved are trained for this role.

Summary

- Identify appropriate selection procedures for different levels in the organisation.
- Identify appropriate and validated practical and psychological tests.
- Identify appropriate and validated physical fitness tests.
- Ensure that attitudinal factors are given equal consideration with those of skill and ability.
- Review interviewing techniques, including training of interviewers.

Training

To maximise employee effectiveness, training is essential. There should be well-designed training programmes for staff at all levels. These will cover various areas.

Ability to do the job Deficiencies in the person/job fit can be minimised by adequate training. Employee training needs should be assessed and organised on appointment and on a regular basis.

The plan should include training programmes for posts and individuals.

Life skills training The workforce is a captive audience for health promotion activity. To be effective the message will need to be repeated in different ways.

Life skills training should be available to all members of staff. Such training allows individuals to understand what are healthy choices for physical and mental health, it enables them to express their views assertively, manage their time effectively and understand the dynamics of relationships. This sort of training is probably most effective in a workshop. Individual training linked to fitness assessment can be useful, particularly in the area of lifestyle, but it is generally an expensive use of specialist resources. Unlike job skills training, participation is voluntary and a variety of initiatives will be needed to maximise the effect of these activities. A detailed health promotion programme should be part of the plan. Further information is to be found in Chapter 8.

Management training

The ability to manage does not come naturally, although some would have us believe that managers are born and not made. There are still many managers who have received no training in the management of people. The skills of good management are easy to define but difficult to practise:

- regular/open communication;
- being able to delegate;
- being able to make decisions;
- fair criticism;
- fair appraisal;
- regular and constructive feedback.

The specialist nature of work today means that most managers have risen up a professional or technical line, their managerial role requirement gradually overtaking their professional or technical role in importance. Well-balanced communication with their own staff may be difficult, leading to either over-indulgence or apparent indifference. Good staff appraisal systems are relatively rare and yet regular constructive appraisal is a significant part of job satisfaction. Tom Peters (1985) has said that the average American manager may take six months to recover from an appraisal. However, feedback on performance is now regarded as an essential part of good management.

There should be no exception to the principle of training in the fundamental elements of good management for anyone in a management position.

The plan should contain a comprehensive management training programme.

Training to manage stress

The plethora of courses and books on stress emphasise the perception of the general workforce that work can produce a set of symptoms (which are well defined elsewhere in this book) to which is attached the diagnosis of stress. Lazarus (1971) defined stress as 'occurring when there are demands on the person which tax or exceed his adjustive resources'. This emphasises the individual nature of stress. Certain factors in the workplace are known to cause stress. Workshops for defined work groups should facilitate the recognition of stress-inducing factors specific to their own work setting and should enable participants to address these issues constructively. Apart from identifying stress symptoms in themselves, it is also important for managers to be able to recognise problems in their staff and have the skills to deal with the issues involved. Factors known to cause stress in the workplace are:

- unsatisfactory working conditions;
- overload;
- role ambiguity;
- role conflict;
- responsibility for people;
- unsatisfactory relationships;
- under- or over-promotion;
- unhealthy organisational structure/culture;
- attitudinal/cultural misfit.

The thrust of such workshops is usually towards the individual. This enables members of the workforce to identify the signs and symptoms of stress in themselves, identify the causes of stress in the workplace and learn stress reduction techniques such as meditation and relaxation. A series of workshops, however, may well highlight organisational problems which need to be addressed.

A programme of workshops will form part of the plan. As with life skills training, attendance at such workshops is voluntary, but a significant percentage of personnel can be reached if workshops are available over a period as colleagues will encourage each other to attend. It should be clear that attendance at a workshop does not identify the individual as being at risk.

A detailed stress management programme is described in Chapter 4.

Training to manage change

No one likes change. The poem below written by Machiavelli in the sixteenth century expresses this very well.

No task is so difficult
To set about,
No leadership so delicate,
No venture so hazardous,
As the attempt to introduce
A new order of things.

Those who change
Find as their adversaries
All those who succeeded well
Under the old order,
And no more than lukewarm
Supporters among those who
Might function under the new.

If major change is envisaged in the workplace, the plan should include a strategy to minimise adverse effects on staff. A successful and trouble-free change process depends on a number of factors:

- workforce involvement;
- adequate funding;
- general security;
- good communications;
- adequate staff support;
- imaginative redeployment;
- good redundancy packages;
- appropriate timescale.

Managers need to know how to manage change; how to change 'no' to 'yes'. Too often major change finds managers beleaguered, with no skills or time resources to negotiate the change successfully.

It is generally believed that change can be more easily effected if:

- there is sufficient discontent with the status quo;
- there is a strong vision of what the new state will be like;
- there is an easy first step;
- the cost is seen as worthwhile (emotional as well as financial).

Summary

- Training plans should be drawn up for each post.
- Training plans should be drawn up for each individual to improve fitness for the post.
- A detailed life skills training programme should be prepared.
- Everyone placed in a management position should routinely receive management training at an early stage.
- A programme of stress management workshops should form part of the plan.
- Where major change is under way or envisaged, the programme should include training managers to manage change, training staff to cope with change, and provision of counselling facilities.

Developing a healthy organisational structure and culture
It is not appropriate in this chapter to describe in any detail theories of good management and variations in organisational structure. In assessing the organisation for the development of the health plan it may become clear that the organisation itself is 'sick', and that efforts directed at employee health cannot succeed without fundamental improvements. Employee-directed initiatives should not be used to plaster over these cracks. Chapter 1 describes these issues in some depth.

Developing health-related policies

Sickness absence
Inevitably, in any organisation some people will become ill. Concern is likely to arise if there is long-term absence or frequent short-term absences. The latter are likely to be related to factors other than illness and require good management. The organisation should have a written sickness absence policy which includes guidance for managers on how to manage both types of absence. This will indicate points at which action should be taken and what this action should be, when medical advice should be sought and what weight should be given to that advice. The role of the employee's general practitioner and that of the occupational

health physician should be clear. The guidance should also indicate the organisation's position on rehabilitation and resettlement. Where there is little flexibility for either of these initiatives, sickness absence is inevitably prolonged. Chapter 6 discusses sickness absence issues.

Smoking
A smoking policy is essential in all organisations. There is a detailed discussion of this in Chapter 7.

Substance misuse
There have been many attempts to quantify the extent of substance misuse (including alcohol) in the workplace. Many organisations have developed sympathetic policies for the management of individuals with this problem.

Unfortunately, most cases remain concealed and form a small but inefficient element in many workforces. An increasingly serious approach is being taken by employers in organisations where there may be a safety or security risk. On recruitment the urine or blood of candidates is screened for drugs or alcohol. In some organisations this is followed by unscheduled testing during employment. It remains doubtful whether such screening is adequate or effective. The substance misuse policy should address this issue with regard to its relevance to the particular organisation. See Chapter 4 for detailed discussion and a model policy.

Equal opportunities
This issue is considered in Chapters 9 and 10 where model policies are provided.

Summary The health plan should include policies on:

- sickness absence;
- smoking;
- substance abuse (including alcohol);
- equal opportunities.

Developing staff support systems

Counselling
Counselling at an early stage can and does reduce poor performance and sickness absence (Reddy 1992). The level of counselling provision will depend very much on the organisation. The original assessment should give some indication of the level of need. In many instances there may be sufficient expertise within the human resource management and the occupational health teams.

Where the need is great, consideration should be given to the appointment of counsellors or to setting up an external counselling provision, using one of the many organisations providing such services. A full employee assistance programme provides counselling for staff on many issues unrelated to work including social and legal problems. Payment is usually on a per capita basis. The cost may not be justified for average need when there is a reasonable level of expertise in-house. The issue of counselling is described in greater detail in Chapter 4.

Support groups

For work groups where the pressure on staff is known to be great, such as health care workers, setting up local support groups may serve to protect employees' health and provide a useful method of defusing developing problems. These groups may provide useful indicators of organisational problems. To be effective such groups should be well constructed and their purpose must be clear. If not properly established they serve little useful purpose.

Summary The health plan should include guidelines on staff support and counselling.

Health promotion

The Health of the Nation (Department of Health 1993) proposed the setting up of a task force to examine and develop activity on health promotion in the workplace. It targeted five areas of particular concern:

- coronary heart disease and stroke;
- cancers;
- mental illness;
- HIV/AIDS and sexual health;
- accidents.

None of these targets can be achieved solely by action in the workplace but all are important to the working population.

The health plan should include consideration of appropriate health screening programmes for this particular workforce and what form health education should take. For further discussion see Chapter 8.

Summary Recommendations for health promotion activities and health screening should be clear.

Implementation and audit

The working party will now be in a position to submit the draft health plan for general discussion. At this stage the cost implications should also be developed. Because the requirements and commitment of individual organ-isations will be varied, it is impossible to give overall guidance on the resources that will be needed.

Following consultation, the working party should draw up the substan-tive plan which will form the organisation's policy on health. The outline of a model health policy is shown on pages 47–9. Some elements will already be in place, others will be possible within a short timescale, yet others may require anything up to five years to develop. Targets should be set for implementation and also for assessment of the results of revised or new activities in areas such as personnel selection and health promotion. It should also be clear who is responsible for implementation and the monitoring of progress.

A model health policy

Introduction

The [name of organisation] recognises the need to protect and promote the health of employees. The provision of a pleasant and safe workplace, where employees will feel enhanced, can contribute significantly to the overall health of employees. The International Labour Organization has defined health as a state of physical, mental and social well-being.

This organisation has a holistic approach to the health of the workforce and has developed health and safety and personnel policies in line with that philosophy. This health plan outlines the actions taken to enhance and protect employees' health.

Improving the person/job fit

Selection procedures

In association with individual line managers the personnel department will develop a full job and personnel specification for each job category.

They will also develop appropriate and practical selection procedures which may include:

- psychological testing;
- assessment of physical fitness;
- aptitude testing;
- attitude/company culture compatibility.

Consideration will be given to the form of interviews and who should participate.

All interviewers will receive training in interviewing techniques and equal opportunities.

Training

Training plans will be developed for each job category and each post holder to improve their ability to carry out the job. The training will be designed to improve the individual's potential.

Everyone placed in a supervisory or managerial position will receive training in general management skills on appointment, if not previously trained.

Where significant change is envisaged, managers will receive training to help them manage the change process. Staff will also receive training in coping with change.

General life skills training will be available for all staff.

Appraisal

Fair appraisal systems will be in place for all employees to provide a mechanism for structured assessment of performance, feedback on that performance, identification of strengths and weaknesses, and a plan for training and development.

Organisational style and structure

Every effort will be made to enhance communication upwards, downwards and sideways, to provide opportunities for personal growth and initiative, and to provide appropriate reward and recognition.

Special health-related policies

The organisation has developed personnel policies to enhance employee health. They include: [for example]

- sickness absence;
- early retirement;
- substance misuse (including alcohol);
- smoking control;
- sexual harassment;
- equal opportunities.

[Detail what is in place.]

Development of staff support

The organisation recognises the need for staff support and counselling. [Detail what is in place.]

Health promotion

In line with the Department of Health publication, *The Health of the Nation*, the organisation has developed a number of health promotion initiatives.

A series of workshops is available to staff on life skills, stress management and surviving change.

Campaigns on, for example, smoking and alcohol abuse are run from time to time.

The organisation is also a member of the Look After Your Heart campaign.

Audit

There is a regular audit of these activities. Managers are responsible for ensuring that staff have access to appropriate areas of this policy. They are also required to put in place measurements of staff health, such as sickness absence, staff turnover and performance.

A checklist of organisational/departmental health

General

1. Organisational culture
   Power ☐
   Role ☐
   Achievement ☐
   Support ☐
   Mixed ☐
2. Communication structure
   Well developed ☐
   Poorly developed ☐
   Various ☐
   General perception   *Good Bad*
   By management   ☐ ☐
   By employees   ☐ ☐
3. Problem-solving arrangements
   Team meetings ☐
   Support groups ☐
   Grievance procedures ☐
   Counselling ☐
   Other ☐
   Detail . . . .
4. Health and safety policy   *Yes*
   Includes mental health ☐
5. Major changes under way   *Yes*
   or envisaged ☐

Personal

1. Role
   Employee relations ☐
   Pastoral ☐
2. Selection procedures
   Well developed
   All staff ☐
   Certain staff ☐
   Detail . . . .
   Personality tests
   All staff ☐
   Certain staff ☐
   Detail . . . .

Aptitude tests
   All staff ☐
   Certain staff ☐
   Detail . . . .
Practical tests
   All staff ☐
   Certain staff ☐
Health screening
   Medical examination
   All staff ☐
   Certain staff ☐
   Nurse screening
   All staff ☐
   Certain staff ☐
   Questionnaire
   All staff ☐
   Certain staff ☐
3. Sickness absence
   Management procedures
   Written Statistics ☐
   Good ☐
   Limited ☐
   Non-existent ☐
   Rates
   High   Average   Low
   ☐   ☐   ☐
4. Health-related policies
   Sickness absence ☐
   Alcohol ☐
   Smoking ☐
   Drugs ☐
   AIDS/HIV ☐
   Employment of disabled ☐
5. Training and development
   For most jobs ☐
   For certain jobs ☐
   Detail . . . .
   Management training ☐

Life skills training
    All staff ☐
    Selected staff ☐
6. Staff representation
    Formal ☐
    Informal ☐
7. Turnover
    Rates

High	Average	Low
☐	☐	☐

Nature of work

1. Jobs with special physical
    standards ☐
    Detail ....
2. Jobs with special
    psychological standards ☐
    Detail ....
3. Jobs with physical hazards ☐
    Detail ....
        Chemical hazards ☐
    Detail ....
        Biological hazards ☐
    Detail ....
4. Shift work ☐

    Detail ....
5. Main contractual arrangements
    Permanent ☐
    Fixed term ☐
    Mixed ☐

Health and safety

1. Safety professional ☐
    Safety consultancy ☐
    Occupational health doctor ☐
    Occupational health nurse ☐
    Non-occupational health
        doctor ☐
    Non-occupational health
        nurse ☐
    Occupational medicine
        consultancy ☐
    Occupational hygienist ☐
    Ergonomist ☐
    Other health professional ☐
2. Safety management structure
    Formal ☐
    Informal ☐
3. Joint safety committee ☐

Part II

Chapter 4
Mental health and illness at work

With increasing well-being, all people become aware, sooner or later, that they have something to protect.

(J. K. Galbraith, *The Affluent Society*, Chapter 8: 111)

As work has become less physical, the psychological health of employees has assumed increasing importance. It has been estimated that 80 million working days are lost each year because of stress-related and mental illness. The case of Walker v. Northumberland County Council (1994) in which Walker, a social worker, succeeded in proving that his employers had not shown sufficient care in preventing his work-related mental illness, suggests that employers may now be at significant risk of litigation. This chapter explores the concept of mental health and the development of mental illness. An attempt is made to overcome misconceptions about the meaning of stress. The causes of stress at work are described, including job overload, role ambiguity, communication problems and change. Guidance is given on the development of a mental health policy. This includes methods of recognising and measuring stressful work situations, techniques for managing stress and change, and relevant personnel policies. The last part of the chapter describes the commonest forms of mental illness and their effect on the individual's ability to work, and suggests ways of managing individuals with mental illness. The chapter concludes with a model policy for mental health and the suggested contents of a stress management workshop.

Introduction

Each year 6 million people in the UK suffer from some sort of psychiatric disorder such as depression or anxiety. Mental health is largely a problem of the working age population and therefore impacts significantly on any business organisation. It has been estimated that 8 million working days are lost each year through drink-related disease and as many as 80 million days through all forms of stress-related and mental illness. In a

Confederation of British Industry (CBI)/Department of Health survey of CBI members in 1991 employers estimated that 30 per cent of sickness absence was related to stress, anxiety and depression. In the same survey 95 per cent of employers considered that the mental health of their employees should be a concern. This concern was not matched by significant action, only 13 per cent of these same companies having developed a policy or programme for promoting mental health in the workforce.

One might pause for a moment to consider the syndrome of stress in the workplace which has been an increasing cause of concern over the last two decades. Is this because the nature of work has become more stressful, or because people's expectations of healthy working conditions have increased? Or is it that life in the latter part of the twentieth century has become generally more pressurised, faster, and hence more demanding? One obvious change in the world of work is from physical to mental activities. The risks to health from exposure to chemical and physical hazards have been significantly reduced through technological advances and effective health and safety legislation. Now a great deal of time in the average workplace is spent relating to electronic information processors. This results in difficulties not experienced when communication is between human beings.

The keyboard operator seems unable to control the pace of work – a new variation of machine-paced performance. Controlled by computer technology, the operator is left without the natural recuperative breaks which occurred in the pre-computer age. The body is overused and exhausted, resulting in problems such as repetitive strain injury and stress.

The role of the manager is now much more complex. Organisational structures are less hierarchical, decisions more participative, objectives more competitive. There is never any time to stand and stare. Add to this the continuing change process and lack of job security, which can be found in almost every industry, and the potential for stress is enormous.

Any plan for mental health in the workplace must address these various causes of stress, suitable health education interventions, facilities for early diagnosis and support, and development of policies enabling rehabilitation and resettlement of those who become mentally ill.

Mental health/ill health

It is probably important at this stage to define clearly what is meant in this chapter by mental health, stress and mental ill health. These form part of a spectrum of disease.

Mental health	*Stress*	*Mental ill health*
physical well-being	sleep disturbance	depression
psychological well-being	fatigue	anxiety state
social well-being	irritability	panic attacks
	anxiety	phobic states
	lack of concentration	substance abuse
	relationship problems	other mental illness
	difficulty in making decisions	physical illness
	feelings of unease	
	substance misuse	
	disturbed eating patterns	

Mental health has been described as a state of psychological well-being which allows the individual to enjoy life and be able to cope with the inevitable problems of living without prolonged or significant change of mood. Well-being is a difficult concept but it certainly includes all the levels of Maslow's hierarchy of needs (Maslow 1970):

- physical needs;
- security;
- social needs;
- ego;
- autonomy;
- self-actualisation.

Stress is a word which has suffered from variations in definition leading to comments such as 'he thrives on stress'. Lazarus (1966) described stress as 'occurring when the individual perceived that the demands of the external situation were beyond his perceived ability to cope with them'. Cooper (1988), in looking at stress in the work situation, proposed that: 'stress is a negatively perceived quality which as a result of inadequate coping with sources of stress has negative mental and physical ill health consequences.'

In other words, a certain amount of pressure enhances performance but there comes a point at which there is too much pressure and stress results, as defined by Lazarus. When put under pressure the body has few physiological responses. These are no different from those available to the cave man or our mammalian ancestors. The available responses prepare the individual to fight or flee. An anxiety-provoking situation stimulates the higher centres of the brain, from which a series of nerve and hormonal connections cause stimulation of the adrenal glands. These glands respond by excreting the catecholamines, adrenaline and noradrenaline, and cortisol into the bloodstream. The effect of these chemicals is to provide the physiological changes that the body may need to confront the problem (the so-called flight/fight response): for example, increases in heart and respiratory rate, concentration of blood in the muscles, increased clotting potential of the blood, and an increase of fats in the bloodstream. None of these changes assists the individual to deal with work overload, a frustrating work situation or a difficult interview with the boss (in fact, a throbbing heart and breathlessness are counter-productive), but that is all we have available.

A continual or recurrent state of stress can eventually lead to anxiety states and depression. Stress has also been shown to be associated with diseases such as cancer and coronary heart disease. Not all mental ill health arises from a gradual deterioration of mental well-being. Although there is disagreement about the influence of the environment on the development of some forms of mental illness, such as depression and schizophrenia, in many individuals these illnesses seem to arise without reference to the general psychosocial environment.

Cause of stress at work

If we now look in more detail at the development of stress symptoms in the workplace there are three obvious components: the mental and physical state of the individual, pressures in the social and domestic environment and in the workplace, and the interplay of all these components.

Individual factors

There is a tremendous variation in the ability of individuals to cope with pressures. Some of the factors which influence the individual's capacity may be permanent, such as genetic factors, social deprivation in childhood, poor parental models. Others may produce transitory vulnerability, such as the occurrence of a number of traumatic life events at any one time, the development of feelings of inadequacy, physical illness and loss of social support.

Factors known to affect the individual's ability to deal successfully with pressures at work are:

- genetic factors;
- poor parental models;
- physical and psychological state;

- personality factors;
- individual behavioural skill repertoire;
- the extent to which the individual feels trapped;
- the extent and intensity of other acute or chronic stressful life events;
- the extent and quality of social support at home and work;
- the extent and outcome of past similar experiences.

Factors in the workplace

A number of factors in the workplace are known to be associated with stress:

- nature of work;
- culture and structure;
- job overload/underload;
- time pressure;
- interpersonal relationships;
- job ambiguity and role conflict;
- career development;
- lack of communication;
- conflict between home/work demands;
- lack of security;
- change.

Nature of the work

Although all work may be potentially stressful, certain types of work are recognised as being more likely to induce stress. This is particularly true of service industries, such as the police force, the ambulance service and health care. In such work there is a continual demand for empathy; the worker may be placed in life-threatening situations and may be frequently exposed to the physical and emotional trauma experienced by others. Other types of work which are inherently stressful are associated with frequent deadlines, short response times and no room for error. These conditions are found in the work of groups such as money brokers and television news teams.

Culture and structure

Not only is the nature of the job a significant factor but also the culture of the organisation. Many organisations where the work undertaken is not intrinsically stressful nevertheless provide employees with an uncomfortable environment where disease is common. Management style, communication structures, objective-setting and appraisal systems may be perceived as unfair or at best quixotic. A significant factor may be a cultural refusal to recognise that stress can be a problem, so there is fear of discrimination if it is admitted.

Job overload

One of the commonest causes of stress at work is overload: too much to do in too little time (quantitive overload) or work which is qualitatively beyond the individual's capacity (qualitative overload). Where there is quantitive overload the problem will be compounded if the employee has little or no control over the load. Machine-paced work has long been recognised as a source of pressure. This applies not only to process workers but also to those working with display screen equipment who are dependent on the timescale which the electronics can achieve. In many organisations the workload of one department is wholly dependent on other departments with little opportunity to control the flow. It is common for an employee to perceive that he is overloaded but be unable to find any solution, or at least any solution not perceived as making him vulnerable to management censure.

Underload, although less common, may cause great anxiety where there are to be job losses or where the individual loses self-esteem because he is not making a worthwhile contribution.

Interpersonal relationships

If there is no relationship of mutual trust and respect between the manager and the member of his team, the subordinate is likely to feel under pressure. The manager may feel equally under pressure when there is a mismatch between formal and actual power, or when a more democratic approach to decisions has been adopted. Unsatisfactory peer group relationships may cause much distress. Scapegoating is not unusual in work groups. This may be difficult for the manager to control and is usually not amenable to outside intervention.

Job ambiguity and role conflict

A common problem for an employee is the lack of a clear job specification. The expectations of the employee may be entirely different from those of the manager or, indeed, the peer group. Often individuals are responsible to more than one manager and may be servicing several work groups. Priorities may be difficult to determine and clarification, when sought, may not be forthcoming.

Lack of communication
It seems almost impossible to achieve good communications in any but the smallest organisations. This may be a minor irritant where the unavailable information will in any case have little impact on the employee. However, in times of change when jobs may seem to be at risk, lack of information and consequent rumour only increase anxiety. Consultation is also important. Communication should be possible upwards, downwards and sideways. There is little point in consultation, however, if there is no discernible impact on management decisions.

Home/work conflict
Extended working hours, unsocial hours and shiftwork all tend to disturb family and social life. It is difficult to say what is a correct balance, although the eight-hour working day does seem to have many credentials. 'Work, rest and play keep the doctor away', 'all work and no play make Jack a dull boy': useful sayings with much real wisdom. Some sort of balance needs to be struck between the compartmentalisation of different aspects of one's life and sharing the workaday world with one's partner. Few organisations approach this problem realistically. The involvement of partners is usually perfunctory.

Change
Cultural anthropologists have found that all human societies evolve in a cultural pattern – a tightly woven system of habits, status, beliefs, traditions and practices. The cultural pattern is a vital stabiliser. Change is often introduced without any consideration of the threat that it may pose to the cultural pattern – which habits; whose status; what beliefs? In such cases resistance is the result.

Change is of such significance that it almost deserves a separate chapter. The continuing change processes which are occurring in all industries have stretched employees' adaptive and coping behaviours. It is difficult to think of an industry which is not undergoing massive change. Much of this is government-led in fields such as health care, education and transport. The aspirations of many organisations to achieve world class and competitive needs are also prime motivators of change. Perhaps a more significant force for change is information technology. Those over 40 can soon feel illiterate. It is difficult to imagine a greater change than that which has occurred in the typing pool. The clattering, noisy, bright environment is now quiet, enclosed, gently illuminated. Human communication is cut to a minimum; in fact, in many cases it hardly needs to occur in the day-to-day work routine. The routine and often inappropriate use of e-mail may reflect the isolation felt by some employees.

In other areas well-developed technical skills are no longer required because sophisticated technology has taken over. Those who were able to accomplish complex tasks requiring manual dexterity and problem-solving strengths may derive little satisfaction from overseeing an electronic system programmed to perform the same task. At the very least, the locus of control has shifted away from the individual.

Holmes and Rahe (1967) have demonstrated clearly (in a scale of some 40 items) that too many changes happening together, giving a total score of 300 or more, may be associated with the development of significant diseases such as cancer and coronary heart disease. An extract from the scale illustrates the relative importance of work-related change:

Life event	*Value*
Death of spouse	100
Fired at work	47
Retirement	45
Business readjustment	39
Change to a different type of work	36
Change in responsibilities at work	29
Trouble with boss	24

To most people change is associated with insecurity either because of a reduction in the number of posts, or because the requirements of the job are subtly changed and the individual may feel deskilled or unskilled and vulnerable. People prefer stability and resist change. Managing the change process successfully is an essential management skill.

Developing a policy for promotion and maintenance of mental health

A full policy may not be necessary in every organisation. Human resource managers need to consider each aspect and decide what is appropriate for their organisation. If the organisational culture and the nature of the work are potentially stressful, or if there is ongoing change in the organisation, training in stress management should be given the same priority as any other form of employee benefit, such as pay for sickness absence.

An example of a policy for mental health is shown at the end of this chapter (see pages 72–4).

Recognising work which is inherently stressful

It is obviously important to recognise that certain types of work are inherently stressful: for example, work which involves dealing with the public at times of stress and trauma, work which involves significant periods away from home, work which takes place in a hazardous environment, work with recurrent deadlines, work where the individual's performance is before the general public, work which involves the ill and dying, and work where the individual may be exposed to physical or mental abuse. It is probably inevitable that individuals who work in such jobs are expected to be able to cope because of some form of self-selection. Expressions of anxiety are seen as signs of failure both by the manager and the individual. Increasing awareness of conditions such as post-traumatic stress syndrome and burnout has made it possible to introduce counselling and other support structures in these areas and, more important, has made it possible for the individual to admit to symptoms. What is offered to employees should be tailored to the organisation's exact needs. It will include opportunities for employees to obtain counselling confidentially and without reference to management. The possibility of resettlement needs to be offered, or at least temporary transfer to less pressurised work without serious career implications.

Recognising organisational cultures and structures which may be unhealthy

An organisation may be described as power, role, support or achievement based. Within these structures the style of management may be equally variable. Although Taylorism has long since been discredited, stick and carrot management can still be found. On the other hand, many organisations are moving towards participative management bringing with it different pressures, particularly for those who are used to a more hierarchical approach.

A significant part of the structure of an organisation is the communication network. Most managements are still seeking a successful communication structure. At times of change, weaknesses in the communication strategy will be easily identified. Such weaknesses can only lead to rumour and distrust.

A caring organisation ensures communication between management and employees. It has in place good appraisal systems with agreed objectives and appropriate recognition and rewards. It allows opportunities for participation in decisions. It has well-trained managers who are fair and consistent in their decisions. Fairness is something that everyone wants but few experience. No amount of effort with employees in training and counselling will prevent mental ill health if their well-being is constantly undermined by an unhealthy management style.

Detecting organisational problems

Early signs of organisational stress are:

- high staff turnover;
- poor morale;
- reduced productivity;
- increased sickness absence;
- customer complaints.

Good sickness absence statistics will help to pinpoint problem areas (see Chapter 6). Questionnaires such as the Occupational Stress Indicator (OSI) may also give an early indication of departmental sickness. The OSI is a computer-based questionnaire which is completed by individual employees. It measures a number of parameters which can be used to counsel the individual but, by combining individual scores, can also be used to give an overall picture of a department. If the problem is not easily identified, an organisational psychologist may be required to undertake further analysis and help to resolve the problems. Sub-scales of the OSI are:

Sources of stress
Factors intrinsic to job
Managerial role
Relationships with other people
Career and achievement
Organisational structure and climate
Home/work interface

General behaviour

Locus of control

Coping mechanisms
Social support
Task strategies
Logic
Home/work relationships
Time management
Involvement

Job satisfaction

Current state of health
Mental
Physical

Managing stress

Employees may be helped to avoid stressful reactions and manage potentially stressful situations in various ways. Stress management workshops
generally help individuals to identify stress symptoms, recognise the cause
and develop strategies to limit the effect of stress-inducing situations and
events. Individual signs associated with stress are:

- reduced performance;
- accident proneness;
- relationship problems;
- lack of concentration;
- impaired judgement;
- ineffectual management;
- reduced creativity;
- slow and poor decision-making;
- sleep disturbance;
- changes in consumption (alcohol, food, tobacco);
- excessive fatigue.

Such workshops appear useful in the prevention of stress, although
there has been little scientific evaluation. They are not particularly useful,
however, in helping the already stressed or ill individual, where one-to-
one methods such as counselling are more effective and acceptable.

In a good mental health plan everyone in the organisation should attend
a workshop over a period of time. This immediately disposes of the idea
that these workshops are for those who are already stressed. An outline
of a workshop is given at the end of this chapter (see page 78). There
are many individuals and organisations who can provide these (see Useful
Addresses, pages 211–15). If there is in-house expertise within the
training or occupational health department, this is likely to be beneficial
since issues particular to the organisation can be more easily addressed.
Perhaps one of the best examples is the programme of workshops run
for a chemical company in the north west of England. All 3000 employees
attended a series of stress management workshops. Unfortunately,
although there was subjective improvement (O'Sullivan 1992), no objective measurements were taken.

Where there are limited resources, time pressure and no in-house expertise, the use of flexible learning packages could be considered. These can be bought off the shelf (see Useful Addresses, pages 211–15) and enable individual employees to work on their own or with a partner to identify what they need and when. Feedback to managers can facilitate organisational change.

Managing change

No one denies that human beings prefer the status quo. Therefore, the introduction of any change in the workplace is likely to be resisted, although there are a number of possible responses:

- anxiety that the society and its culture will change;
- belief that it can only happen to others;
- loss of confidence in responsible bodies;
- fear of loss of income;
- fear of the unknown;
- belief that it could work out for the best.

To accomplish a successful and untraumatic change, or at least one with the minimum number of casualties, strategies should be put in place at an early stage. Successful change management should include:

- good communications;
- realistic timing;
- clarification of issues;
- clarification of choices;
- counselling support.

In addition, managers need to develop skills in managing change. This includes not only managing technological change but understanding the social consequences. Employees need to understand how they are responding to the change process. Where resettlement or redundancy is inevitable, individual expert counselling should be available.

Employee support

In many organisations, not all small, support for the troubled individual is not seen as an employer's responsibility. However, as we have seen from the CBI survey (*Working for Your Health* 1993), a significant percentage of employers are concerned about the mental health of their staff.

Any such supportive activity is likely to be difficult for line managers to perform except in basic terms. Managers should, however, be encouraged and trained to address such problems as they are likely to be the initial point of contact. Having established that there is a problem, managers may not find it easy to procure help for the employee. The traditional pastoral role of personnel departments now seems to be largely replaced by employment law and industrial relations activities. Access to occupational health experts is also limited. Where there is an occupational health service this should provide a substantial counselling input. Organisations with a wide geographical spread and perhaps fewer than 100 staff on each site are not well placed to provide individual support. Employee assistance programmes, originally developed in North America to meet the counselling needs of those who are drug or alcohol dependent, now usually provide general counselling support. And this may be an appropriate support system for widely dispersed organisations. It may also be chosen where there is particular sensitivity or paranoia about any internal intervention. Payment is usually on a per capita basis and requires a considerable financial commitment from the employer with no real feedback or check on efficacy.

Personnel policies

Personnel policies which enshrine a caring response to mental ill health should enable employees to reveal problems at an early stage, thus preventing the development of serious mental illness. In addition, good resettlement and rehabilitation programmes may lessen the loss of key staff. Policies on sickness absence, alcohol and other substance misuse are particularly important.

Mental illness in the workplace

As already suggested, some mental illness may be a result of work or social environment, or of individual vulnerability. Either way, it will be necessary to manage employees who have developed significant mental illness. Mental illness may result in bizarre behaviour which is frightening to the observer and may sometimes be associated with danger. Where behaviour was bizarre, it may be difficult for colleagues to accept the return of the ill person to the workplace on recovery. There is a lingering belief that, unlike physical illness, mental illness is something that one has brought on oneself and is controllable. When those who have been mentally ill return to the workplace, their colleagues may find it difficult to treat them as normal people. The anxieties of colleagues may be reduced if a degree of openness about the condition is possible.

Most mental illness is of a relatively minor nature and may well not recur. Illness precipitated by an event such as a bereavement is unlikely to result in long-term problems. Similarly, where a stress illness such as an anxiety state, a panic attack or a phobia is associated with a particular situation at work or at home, it is unlikely to recur if the precipitating factor can be removed or if the individual learns techniques to control the problem.

The three most serious mental illnesses that employees may suffer from are schizophrenia, mania and depression. The most common condition likely to be encountered is some form of anxiety state. The possible effects on work of these conditions and the treatments used are discussed briefly below.

Alcohol and drug misuse are also discussed and an example of an alcohol policy is shown at the end of the chapter (see pages 76–7).

Schizophrenic psychosis

In this illness there is a fundamental disturbance of personality associated sometimes with hallucinations and delusions.

Work effects

Work problems may result from either an acute episode of disease or a chronic illness state. Chronic schizophrenia may be well controlled and not incompatible with work. As a result of the disease and the drugs used to control it, the individual may remain somewhat withdrawn and may not respond well to pressures at work associated with deadlines and changes in the work situation. Often schizophrenics are employed in jobs below their intellectual capacity but in keeping with their tolerance of pressure.

In acute cases fellow workers may notice that the individual is withdrawn. There may be periods of normal productivity and others when nothing is done. There may be unusual actions and inappropriate responses. It may become obvious that the person is suffering from delusions. This may, of course, be very disturbing to colleagues. Every attempt should be made to ensure that a doctor is consulted. Following treatment for an acute episode a return to normal behaviour may be rapidly achieved.

Manic-depressive psychosis

This disorder is associated with serious disturbances of mood such as depression, excitement and elation. Recurrent depression is more common than recurrent mania. Manic phases are associated with excessive activity, feelings of elation and garrulousness. Depressive phases are associated with sadness, loss of energy and concentration, and sleep disturbance (usually early morning waking).

Work effects

Pre-employment considerations will be the history of frequency and severity of attacks. It is not uncommon for individuals to have only one attack of depression and never experience any manic manifestations. A history of a severe depression with a full recovery and a reduction in or completed treatment may suggest a good prognosis. In the case of manic attacks complete control is often attained by long-term medication. Lithium is the drug commonly used to control this condition. It has no side effects which affect the ability to work. It is usual for the individual to recover fully from an attack and in many cases there is no recurrence.

Anxiety states

An anxiety state is one where there are various physical and psychological signs of anxiety unrelated to any realistic danger. It may present as a panic attack or a more chronic distressed state. Symptoms such as sleeplessness, palpitations and phobic ideas are common. A variety of physical symptoms may be associated with anxiety states and these may interfere with the correct diagnosis.

Work effects

Anxiety at work may develop slowly with a gradual deterioration in performance. Long-term sickness absence and long-term medication are not usually required. The individual may need counselling support and the removal of any precipitating factors before rehabilitation can be completed. In some cases a chronic state may develop where unreasonable anxieties and loss of self-esteem persist.

Alcohol misuse

Problem drinking at work may be the result of established alcohol dependency which will need professional treatment, but may also be a behavioural problem which can be controlled by the individual. It is not always easy for the manager to make a decision as to which of these situations prevails. Drink problems are common in the workplace. The financial cost to industry has been estimated at over £1300 million per annum (Royal College of Physicians 1987). It is estimated that 8 per cent of the population are heavy drinkers, 2 per cent are problem drinkers and 0.4 per cent are alcohol dependent.

Work effects
Identifying someone with an alcohol problem at work can be extremely difficult. Management of the individual, when alcohol abuse has been identified, is recognised as being fraught with difficulties. Those with a drinking problem may be at increased risk of accidents and frequently absent, particularly following rest days such as weekends. Impaired efficiency in the afternoons and general irritability may also point to this problem. Some organisations now undertake spot checks of blood for alcohol levels at the pre-employment stage; such checks are usually only used where there are public safety issues associated with employment.

A policy on alcohol misuse is essential to any consistent management of the problem. A sample policy is given at the end of this chapter (see pages 75–7). Such a policy should contain the following elements:

- restriction of alcohol on the premises;
- a structured approach to rehabilitation;
- health education;
- training of managers on recognition and management of alcohol misuse.

The co-operation of the affected individual is more likely to be forthcoming if he or she is reassured about a caring approach and that the condition is seen as an illness. The alcohol policy should clearly state the management approach to employment issues but make it obvious that the individual is expected to co-operate. Education on issues around drinking and the training of managers to deal with such cases should also form part of the strategy. Up to 50 per cent success has been reported in some areas.

Drug misuse

This has been defined as the taking of drugs to the detriment of the person's health and performance. Drugs commonly misused are heroin and cocaine, amphetamines, barbiturates and other stimulants such as

LSD. The abuse of drugs appears to be increasing. A recent study of drug use among teenagers in Manchester showed that 70 per cent had experimented with ecstasy and 41 per cent with cannabis (Edwards *et al.* 1988).

Work effects

Acute effects may severely affect work performance. As with alcohol abuse, it may be difficult to detect and deal with the problem in the work setting.

However, there are important legal differences since the non-medical use of controlled drugs such as heroin and cocaine is illegal (the Misuse of Drugs Act 1971). An employer is required to report any trafficking in drugs occurring on the premises and failure to do so may lead to prosecution.

Where there appears to be a special problem within an organisation, or where there are special safety issues, a drug misuse policy should be agreed.

Some organisations now screen potential employees for drug misuse. However, as in the case of alcohol this is only used at present where there are safety implications.

Conclusion

Major psychiatric illness is not a significant problem in the workplace. More sickness absence, disruption of work and generally inappropriate behaviour result from stress-related disorders such as anxiety states, reactive depressions and stress-induced physical ill health. Management needs to address the causes of stress and, where a high level of pressure is inevitable, adequate staff training and support should be provided.

A policy for mental health

Introduction

—— is committed to providing and maintaining the health of all staff. The company's policy on health and safety makes specific reference to mental health and this policy puts into effect measures designed to maintain the mental well-being of staff by addressing the known causes of stress at work. Stress is defined as a situation or condition where the pressures experienced by an individual exceed that individual's ability to cope.

Causes of stress at work

Causes of stress at work have been identified as:

- job overload – quantitative or qualitative;
- poor person/job fit;
- role conflict;
- lack of role definition;
- interpersonal relationships;
- communication problems;
- change;
- monotonous tasks;
- lack of opportunity for personal development;
- perception that job is not important;
- poor working conditions;
- unsatisfactory hours of work.

The organisation has in place various initiatives to reduce stress.

Organisation structure and outline

The organisation has a support-based culture with a network system of management. This enhances job satisfaction and encourages independent initiative.

Communications

The organisation has in place clear lines of communication to facilitate the passage of information from management to staff, from staff to management, and between different but interdependent work groups.

Managers must ensure that a formal meeting of all staff for the exchange of information is held at least monthly. They should also allow time for informal meetings on a regular basis to exchange ideas and attitudes.

Role definition and job description

Every post shall have a clear job description with clear reporting lines and responsibility.

Selection of staff

The required background and experience for each post will be fully described. Appointments are made without reference to sex, race or disability. Every effort is made to fit the person to the job.

Training of staff

General

The required training on recruitment and as required for the skills development of the individual will be regularly defined and approved. Where the annual appraisal system has identified weaknesses every effort will be made through training to assist the individual to repair these.

Manager

All managers will receive full management training at an early stage in their appointment, if they have not already received it.

Staff appraisal

All staff participate in the well-structured annual appraisal scheme which provides feedback on performance, gives an opportunity for individuals to identify their own areas of concern, and identifies weaknesses and training requirements.

Health education

The organisation provides life skills and stress management training on a regular basis. Information on these subjects and on areas such as substance misuse are available from the occupational health department.

Sickness absence

The organisation's sickness absence policy does not differentiate between physical and psychological illness. Sick pay, rehabilitation and retirement procedures are identical.

Counselling

Confidential counselling is available to all staff through the occupational health department. Individuals are encouraged to seek assistance.

Problem-solving

Every department has a well-established system for problem identification and solution. If this does not appear to be successful, individuals are encouraged to discuss the problem with their personnel officer.

Change

The organisation recognises that change may be particularly stressful. To facilitate change the following procedures are followed. All staff likely to be affected are given:

- full information as soon as possible;
- the opportunity to discuss likely personal problems;
- career guidance;
- counselling support if required.

Rest and holiday breaks

All jobs are organised to allow reasonable rest breaks and holidays. Staff are encouraged to take these breaks. It is the responsibility of managers to ensure that breaks are taken.

Shiftwork

Where shiftwork is necessary it is designed to cause the least possible detriment to staff. If a change of shift hours is proposed staff have the opportunity to accept or reject the proposal.

Conclusion

As part of the annual health and safety audit, the performance of each department in respect of the above measures is recorded. Managers are expected to ensure that all measures are implemented.

A policy for dealing with misuse of alcohol

1. Introduction

1.1 Alcohol misuse is widespread in the community and this organisation recognises that there will be employees who have alcohol-related problems. It believes that all employees should be assured that if they are identified as having an alcohol-related problem which adversely affects their work, they will be offered assistance in obtaining advice and whatever other help is considered necessary after assessment by the occupational health department.

1.2 Alcohol misuse can result in higher levels of absenteeism and accidents, a decrease in activity and an overall deterioration in work performance and relationships at work.

1.3 At all times it is important that staff are functioning at their optimum level. There is considerable evidence that even one alcoholic drink can impair performance.

1.4 Any indication of alcohol consumption by an employee may affect the customers' confidence in the organisation.

2. The policy

2.1 Alcohol must not be consumed by employees immediately before coming on duty, during the working period (including breaktimes) and while on call.

2.2 The organisation will set up health promotion campaigns for staff to make them fully aware of the risks associated with excessive drinking.

2.3 The organisation will provide assistance and support to staff to help them overcome a drinking problem.

2.4 The organisation will maintain the strictest confidentiality within the limits of what is practicable and within the law.

The attached procedure for dealing with misuse of alcohol must be followed.

A procedure for dealing with misuse of alcohol

1.   Aims of this procedure

1.1   To provide assistance to employees who suspect or know that they have a drink problem.

1.2   To assist managers in dealing with alcohol-related problems in a fair and equitable manner.

1.3   To advise managers and employees of the implications of drinking at work or during working hours.

2.   Identifying an alcohol problem

2.1   In addition to being aware that an individual may be drinking an excessive amount of alcohol, other warning signs may include:

(a) impaired performance;
(b) lateness and absenteeism;
(c) irritability, tremor, slurred speech, impaired concentration, memory lapses, deterioration in personal standards and dress;
(d) bouts of anxiety or depression.

3.   Stages to be followed

3.1   Stage 1 – Exploratory talks

If a manager or supervisor identifies or suspects that there is an alcohol-related problem, he or she should discuss the matter with a personnel officer or an occupational health officer and then separately with the individual.

3.2   Stage 2 – Advice from occupational health staff

At this stage, the employee should be offered the assistance of the occupational health staff who are able to offer support and advise on outside agencies if appropriate.

If the employee accepts referral to the occupational health department, no further action should be taken until the occupational health physician advises the personnel officer and the manager whether there is an alcohol-related problem. If no alcohol-related problem is identified, the normal disciplinary/sickness procedures should be followed.

3.3 Stage 3 – Refusal of assistance

The decision whether or not to accept treatment or advice has to be that of the individual. However, if an employee declines to receive treatment and his or her standard of performance or level of attendance remains unacceptable, he or she may be subject to the disciplinary procedure.

3.4 Stage 4 – Acceptance of assistance

Assistance and referral for treatment will normally be offered to all employees, although there may be occasions when this is not appropriate; for example, when treatment has not been successful on a previous occasion or where the consequences of drinking have been too serious.

If an alcohol-related problem is identified by occupational health staff, the occupational health department will advise the manager of the likelihood of successful intervention, whether absence from work would be appropriate, and what co-operation and support are required to facilitate recovery, taking into account the employee's duties and the service being provided by the department.

If an alcohol-related problem is identified and time off work is required, the period should be deemed as sickness absence and paid accordingly. The rules regarding certification equally apply. If the employee co-operates and work performance returns to an acceptable level, no further action is required, but the situation should be carefully monitored by both the manager and the occupational health department.

If, during or after treatment, a relapse occurs, the manager should consider the merits of the case, discuss the situation again with the occupational health department and the personnel officer and decide whether to offer further opportunities to accept help. Disciplinary action may be considered at this stage.

3.5 Stage 5 – Dismissal

Dismissal in accordance with the organisation's disciplinary procedure may be the only course of action left open to management.

Contents of stress management workshop

General

Didactic	What is stress?
	Physiological aspects
Brainstorm and discussion	Symptoms of stress
Didactic	Relationship between stress and disease
Brainstorm and discussion	Causes of stress at work
Didactic	Dealing with stress at work

- assertiveness
- time management
- communication skills
- the change process

Change

Individual work	How I have experienced change
	Life changes scale

Individual characteristics

Questionnaires and discussion	Type A/B personality
	Coping skills
	Exercise programmes

Relaxation techniques

Didactic and experiential	Relaxation techniques
	Meditation
	Visualisation

Chapter 5
AIDS and employment

As if to shield it
From the pains that will go through me
As if hands were not enough
To hold an avalanche off.

(Tom Gunn)

AIDS still generates considerable fear and prejudice in the workplace, but much has been achieved by the development of personnel policies demonstrating the intention of employers to deal fairly with cases, should they arise. This chapter looks briefly at the history and the nature of AIDS with particular reference to causation and transmission. Estimates of the likelihood of infection from different types of exposure are given. Although there is no medical, ethical or legal reason why an employee with AIDS should be treated differently from anyone else with a life-threatening illness, it has in practice been sensible to develop a specific AIDS policy because of workforce anxiety. Details are given of the current situation with regard to AIDS policies in various organisations and the rationale for developing a policy. The possible contents of the policy are discussed including the pros and cons of HIV testing. The chapter concludes with a section on legal considerations. A model AIDS policy is appended.

Introduction

The fear and prejudice generated by the appearance of a 'new' disease, Acquired Immune Deficiency Syndrome (AIDS), were unprecedented. Public awareness developed rapidly in the mid-1980s when it became clear that this was a wasting disease which affected young people in particular and was associated with debilitating infection and malignancies. To compound the anxieties generated by this apparently inevitably fatal disease, the cause was unknown. Identification of a relationship between the disease and a virus now called the Human Immuno-deficiency Virus

(HIV) offered hope of an eventual cure, but this possibility at present seems as remote as ever.

Clinical considerations

AIDS was first identified in 1981 in male homosexuals in the USA. It presented as a wasting disease, associated with multiple infections, an unusual skin cancer known as Kaposis sarcoma and mental deterioration. Death was often rapid. At this stage no cause for the disease had been identified. By 1984 it had been shown that antibodies to a newly ident- ified virus (HIV) could be found in patients with AIDS. Infection with this virus caused the individual's immune system to produce chemicals (antibodies) in an attempt to counteract the infection. These antibodies remain in the blood and can usually be detected two to three months after the infection. Individuals in whom these antibodies can be detected are defined as HIV antibody positive. HIV antibody positive individuals generally develop full-blown AIDS within ten years, although the relationship between the HIV antibody positive state and AIDS still remains unclear and one scientific group disputes that it is causal. Whether the individual is HIV antibody positive or has AIDS, he or she is potentially infectious.

A vaccine against the disease has still to be developed. Similar uncert- ainty surrounds the effectiveness of the only drug, Zidovudine (AZT), that has been used to any extent in the treatment of those who are HIV antibody positive or who have developed AIDS. Each new study into its effectiveness seems to contradict the one before.

Studies of the incidence of AIDS have fortunately demonstrated how the disease is spread and it is clear that the blood of an infected individual has to enter the bloodstream of another for infection to occur. Therefore, routes of infection are:

- male homosexual practices;
- blood transfusion;
- mother to baby;
- intravenous drug abuse;
- certain heterosexual practices.

This knowledge has made it possible to prevent the spread of the disease by targeting at-risk groups with health information. There is no possibility of contracting the disease from touch or use of food and drink utensils. Neither is there any likelihood of infection as a result of eating 'contaminated' food.

Predictions about the number of cases in the UK have varied enormously. Whether because of effective health education or the continuing lack of confirmation about the disease process, the expected number of cases for the 1990s has been continually reduced. The latest estimate from the Communicable Diseases Research Centre is of 2019–2720 new AIDS cases in England and Wales in 1995. At the end of 1997 it is estimated that there will be 4190 AIDS cases alive and an additional 4205 cases of other forms of disease related to HIV infection. It is estimated that at the end of 1991 there were 23,400 HIV-infected individuals of whom 13,900 were homosexuals, 2000 were drug injectors and 6500 were heterosexuals (CDR 1987).

In 1991 a survey was undertaken using serum from anonymous subjects. Although at that time only 1500 cases of HIV infection were known to have been the result of heterosexual contact, the results from the anonymous survey suggested that a more accurate figure would be between 7000 and 8000 (Day 1993). These are not, of course, large numbers set against, for example, 15,000 deaths from breast cancer and deaths from coronary heart disease each year. However, the greatest proportion occur in the working population and for the reasons already given disproportionate anxiety is caused in the workplace.

Some attempt has been made to assess risk of infection. Lifeshield Foundation (1990) figures show:

- *in a sexual relationship* – chances of a new partner being HIV positive: 1 in 300
- *sharing needles* – chances of needle being infected: 1 in 8
- *health care worker giving injection* – chances of patient being infected: 1 in 500

Developing a policy for HIV/AIDS

There is no medical, legal or ethical reason why an employee with AIDS should be treated differently from anyone else. However, most organisations have decided that a specific policy is called for. The need to develop a policy on this topic has enabled organisations to review their policies and practices relating to general disability and disease, and perhaps to develop a more pragmatic and humanitarian approach. To be successful the policy needs to have certain features.

- The policy should be developed before the problem arises. Experience has shown that it takes time to develop a satisfactory policy. Trying to work out strategies when a case presents itself is more likely to produce injustice and inconsistencies.

- The policy should be developed in consultation with staff. This process can in itself be educative and dispel anxieties.
- The policy should be appropriate to the particular workplace. Although certain aspects will be similar in most workplaces, such as the advice to first aiders, clearly there are workplaces with greater potential risk, such as health care and the prison service.
- Management commitment must be at the highest level.
- The policy must be communicated clearly.
- The policy must be accompanied by counselling provision.

Who has policies?

The *Health and Safety Information Bulletin* undertook a survey in 1991 of the UK's largest employers. The aim of the survey was to discover the response of these employers to the possible problems of AIDS and employment. The key findings of the survey were:

- Seventy per cent of those surveyed have a policy on AIDS. Public sector employers are more likely to have a policy than private sector employers.
- Larger organisations are more likely to have a policy regardless of sector.
- Responsibility for operating the policy lies with the personnel department.
- Organisations without policies are not always less advanced in their approach to AIDS.
- Under 25 per cent of private companies provide any kind of information or education, whereas the practice is common in the public sector.
- There is widespread opposition to the introduction of HIV antibody tests to screen job applicants/employees.
- Most organisations claim that they do not discriminate against those who are HIV positive but symptomless.
- Most organisations will redeploy employees with AIDS if they so request.
- Most organisations will eventually retire an employee with AIDS on the grounds of ill health.

The survey showed that most policies had been developed between 1985 and 1987 when there was maximum publicity and employee concern. For most employers there has been little incentive to develop a policy since then, with the exception of organisations employing health care workers who are affected by Department of Health guidelines (1983).

Reasons for developing a policy

There are some arguments against developing a policy specifically for AIDS which, it is said, should not be seen as different from other debilitating and life-threatening diseases. However, it would be unrealistic not to recognise that the condition still attracts fears and prejudices which can be disruptive and may be better addressed by a specific policy.

Contents of the policy

A sample policy is given at the end of this chapter (pages 89–94).

Recruitment

This has been one of the major issues for personnel directors. General recruitment health screening by questionnaire, interview or medical examination is widespread and has five main purposes:

- to assess fitness to undertake the work;
- to ensure safety in the conduct of the work;
- to avoid sickness absence burdens;
- to avoid training personnel who may be lost through early retirement;
- to assess fitness for entry into the superannuation scheme.

Most organisations do not now require pre-employment medical examinations except when there are major physical requirements or potential safety problems. It has been clearly demonstrated that such medicals give little or no indication of medium- and long-term sickness and sickness absence. There is no justification for HIV antibody testing or for a direct question on employment. Indeed, if it were carried out on the basis of sex or race it could be considered to be illegally discriminating. There are several reasons why testing for HIV antibodies and direct questioning are neither justified nor useful:

- The natural history of the disease, as described earlier, indicates that there is no set pattern for the development of AIDS in those who are HIV antibody positive.
- An individual may have no antibodies at the time of the test but, of course, may be exposed to and develop antibodies subsequently. Regular testing of all employees is clearly inappropriate and costly.
- Those who know that they are, or believe that they might be, HIV antibody positive will avoid situations where a statement or test is required and will avoid being tested, with consequent increased chances of spreading the infection.

In only two areas is HIV antibody testing currently carried out on employees or potential employees:

- *Airline crews* – this has been justified on the basis that their lifestyle is irregular and this is known to potentiate the development of AIDS in an HIV antibody positive individual; and because they are required to have frequent vaccinations and immunisations which may be impacting on an already damaged immune system.
- *Staff on overseas postings* – some countries require a clearance certificate for entry.

Health care workers are in a different category from other employees. Such workers may themselves be at risk from HIV-infected patients, although up to the end of 1992 only 148 cases of occupationally acquired HIV infection had been reported worldwide. Health care workers are required by their professional bodies to report the fact if they know themselves to be HIV antibody positive or if they have reason to suspect that they may be. There are three clear reasons for this:

- An injury to them while performing invasive procedures may result in their blood contaminating the patient's blood.
- AIDS may be associated with the development of various infectious conditions which can be passed on to patients made vulnerable by their own disease.
- A significant percentage of individuals with AIDS experience intellectual deterioration.

Incapacity and sickness absence

In general, individuals who are HIV antibody positive need counselling and support. This is hardly surprising when they have to cope with a potentially fatal disease for which there is no known cure. Added to this they have to deal with prejudice and feelings of stigmatisation. Although there are exceptions, most employees who are HIV antibody positive or have AIDS do not want to share this information with work colleagues. They may confide in their manager, or the personnel or occupational health department, and they should be able to do so with a guarantee of confidentiality.

There may be reasons why a limited number of other people need to know:

- when there is disciplinary action under way concerning frequent sickness absence;
- where they may be putting, or have put, other people at risk, e.g. health care workers;

- when deteriorating health requires redeployment;
- when deteriorating health leads to retirement on the grounds of ill health.

The existence of an AIDS policy which guarantees confidentiality within these limits will help to ease the situation. The individual who first receives the confidence should obtain written consent from the sufferer to inform those who 'need to know'.

Deteriorating health and performance should be dealt with as in the case of any other chronic disease. Clearly, a point may be reached where the level of attendance or performance is unacceptable, and consideration must be given to redeployment to less onerous duties or hours, or ill health retirement.

The employer does not have right of access to medical information as is clearly demonstrated in the Access to Medical Reports Act 1988. In the absence of any information as to the cause of unacceptable absence or performance, the employer is entitled to follow the normal disciplinary procedures.

Training and education
It is essential that the policy covers the training and education of staff. This should include reinforcement of the principles stated in the policy, the facts about AIDS including epidemiology and mode of infection, and advice on reducing the risk of contracting the disease both in the social arena and at work. The education should also reassure employees that there is no risk of contracting AIDS from using crockery, glasses, towels etc. which have been used by an HIV antibody positive colleague.

First aid
Trained first aiders should be reassured about the risk of infection and informed clearly how the disease can be contracted. They should already know how to avoid direct contact with blood and body fluids because of the risk of hepatitis B virus infection.

Precautions include:

- covering their own abrasions with waterproof dressings;
- using disposable plastic gloves and apron when clearing up blood and body fluids;
- using a mouthpiece for mouth-to-mouth resuscitation.

They should also be reassured that the risk from mouth-to-mouth resuscitation is negligible and that this should not be withheld if a mouthpiece is unavailable.

Staff who travel overseas

Avoiding the risk

The greatest risk to overseas travellers is sexual transmission. In eastern and central Africa and in much of the Caribbean and South America, the main route of transmission is heterosexual intercourse. In some of these areas the chances of a partner being infected is one in five. Travellers should be advised to abstain or use safer sex techniques and condoms.

Injecting drug users will also be at risk from contaminated syringes and needles. In some European and American cities up to 80 per cent of drug users are probably infected.

It cannot be assumed that all blood used for transfusion has been tested for HIV. In a country where there is a high incidence of HIV infection, travellers should avoid blood transfusion unless it is essential to preserve life or unless there is convincing evidence that the blood has been screened.

Medical treatment is potentially hazardous if syringes need to be used. Frequent travellers should be provided with a small pack of syringes and needles.

Travellers should be advised to avoid any procedures which puncture the skin, such as tattooing.

HIV antibody testing

Some countries insist on a certificate of clearance; anyone intending to live or work in that country must have a certificate stating that he or she is HIV antibody free. Admission will not be allowed without this.

Insurance

Some insurance companies may refuse life insurance to those who will be working in countries with a high incidence of HIV infection. Where insurers do provide cover they may require a higher premium.

Legal considerations

Recruitment of individuals with HIV/AIDS

Employers have the right to decide whom they wish to employ but they must not discriminate directly or indirectly on the grounds of race or sex: for example, requiring information on HIV infection either by question-naire or blood testing only from men.

Dismissal of employees with HIV/AIDS

Any full-time employee with two years' service is protected by the Trade Union Reform and Employment Rights Act 1993.

Testing for HIV antibodies

Implicit in every contract of employment is that employees will obey reasonable instructions, but only in exceptional cases would HIV testing be justified as relevant to the employee's capability to do the job. If the employee is pressured to undergo the test, he or she may have a claim for constructive dismissal. If male, he may be able to claim unlawful indirect discrimination.

An example of an unreasonable instruction was seen in the catering industry where homosexual male chefs were required to undergo a blood test and then moved to non-food handling work. There is no medical reason why those who are HIV antibody positive should be removed from food handling.

Hostility from colleagues

Problems arising from the fear or hostility of colleagues were common in the early days of public recognition of the disease. The level of education about AIDS now makes such an occurrence unlikely.

If a colleague asks to be moved away from the infected employee and refuses to work near him or her, every effort should be made to reassure by providing the facts about AIDS and infection. If the individual persists in the request or refuses to work, the employer has every right to dismiss the protester (*UCATT v. Brain* 1981).

More commonly, colleagues will put pressure on the employer to dismiss or move the person with AIDS. If dismissal results this will normally be deemed to have been unfair. If the employer has been faced with industrial action the dismissal will still be perceived as unfair.

The development of a policy on AIDS, as previously described, should prevent these problems arising among the workforce.

Customer pressure

If there is customer pressure to dismiss an employee, as in the case of pressure from colleagues, the employer must endeavour to allay the customer's fears. If this fails and the economic threat is significant, the dismissal may be seen as fair.

Conclusion

Much of the public concern about HIV antibody positive and AIDS patients has been reduced by successful government and other educational programmes.

When an infected individual presents in the workplace, the situation is likely to be handled unemotionally and fairly if an AIDS policy is in place. There should be an assurance of confidentiality, no discrimination and a guarantee that procedures to deal with redeployment and retirement will be the same for all employees. It should also be clear that victimisation and harassment will not be tolerated.

A sample policy on AIDS/HIV-infected health care workers

Section I Management of infected health care workers

I.   Introduction

1.1   This policy reflects the need to protect patients and provide safeguards for the confidentiality and employment rights of HIV-infected health care workers. It is based on the guidance given by the Expert Advisory Group on AIDS in the Department of Health Document (1991) *AIDS-HIV Infected Health Care Workers. Guidance on the Management of Infected Health Care Workers.*

1.2   Human Immuno-deficiency Viruses (HIV), the aetiological agents of Acquired Immune Deficiency Syndrome (AIDS), may persist in infected individuals and be transmitted to others in contact with their blood or secretions. Most transmission occurs sexually, perinatally or by transfer of contaminated blood.

2.   Estimating the risk

2.1   The number of HIV-infected health care workers is unknown. In the USA, 5.4 per cent of patients suffering from AIDS are health care workers. Since they make up 5.7 per cent of the workforce, it seems that they are no more likely to be HIV positive than the general population.

2.2   The risk of acquiring HIV from an infected health care worker is extremely small and has been estimated by the Centre for Disease Control as less than 24 per million. Prospective studies in the USA and elsewhere on patients undergoing invasive surgery or dental treatment revealed a seroconversion rate of 0.06 per cent. This represents a negligible risk compared with a 20–30 per cent risk of seroconversion from needlestick injury involving hepatitis B positive material. Further studies since 1982 have examined retrospectively the possibility of transmission from HIV positive surgeons. Serological testing of over 1000 patients operated on by these surgeons has revealed no cases of HIV transmission.

2.3   The evidence available indicates that there is a far greater risk of transmission of HIV from infected patients to health care workers than from workers to patients. Up to December 1992, there had been 148 reported cases worldwide of health care workers infected with HIV through contact with their patients.

3.   General principles of infection control

3.1   Provided that routine infection control measures are taken (Safe Practices and Techniques with Blood and Body Fluids Control of Infection Policy), the circumstances in which HIV could be transmitted from a health care worker to a patient are restricted to exposure-prone invasive procedures in which injury to the health care worker could result in the worker's blood contaminating the patient's open tissue.

4.   Exposure-prone invasive procedures

4.1   Exposure-prone invasive procedures are defined as: surgical entry into tissues, cavities or organs; repair of major traumatic injuries; cardiac catheterisation and angiography; manipulation, cutting or removal of any oral or peri-oral tissues, including tooth structure, during which bleeding may occur; vaginal or caesarian deliveries or other obstetric procedures during which sharp instruments are used.

4.2   The risk of injury to the health care worker depends on a variety of factors which include the type of procedure, the skill of the operator, the circumstances of the operation and the physical condition of the patient. Examples of procedures where infection may be transmitted are those in which hands may be in contact with sharp instruments or sharp tissues (spicules of bone or teeth) inside a patient's body cavity or open wound, particularly when the hands are not completely visible.

4.3   Such procedures should not be performed by HIV-infected health care workers.

4.4   The UK Advisory Panel on HIV-infected health care workers should be consulted where there is doubt about whether an individual's activities need to be restricted. The Panel has been established to provide advice to the occupational physician, or other physician responsible for an infected health care worker, on the activities that such a person may safely pursue.

5.   Action by the infected individual

5.1   The professional codes of conduct for and ethical responsibilities of doctors, nurses and other health care staff have been defined by the relevant professional bodies. All health care workers have an overriding ethical duty to protect the health and safety of their patients. Those who believe that they may have been exposed to infection with HIV in whatever circumstances must seek medical advice and diagnostic HIV antibody testing if applicable. Those who are infected must seek appropriate medical and occupational advice to ensure that they pose no risk to patients.

5.2   Infected health care workers who perform exposure-prone invasive surgical procedures must obtain occupational advice on the need to modify or restrict their working practices. Initially, such advice may be sought from their own physician, but arrangements should be made to seek advice from the consultant in occupational medicine.

6.   Management of HIV antibody positive staff

6.1   In order to minimise the scope of ambiguity and conflict of interest, it is recommended that all matters arising from and relating to the employment of HIV-infected health care workers are co-ordinated through the consultant in occupational medicine.

6.2   Further course of action will depend on the nature of work undertaken by the member of staff with particular emphasis on those involved in invasive procedures.

6.3   If specialist counselling has not already been received, the consultant in occupational medicine will immediately arrange this.

6.4 Staff involved in invasive procedures

6.4.1 The consultant in occupational medicine will discuss with the individual any alteration in work activity which may be necessary. Those who are involved in invasive procedures must cease these activities immediately.

6.4.2 With the consent of the individual, the head of department, or anyone else whom the staff member wishes, may be brought into the discussions to facilitate modification of duties.

6.4.3 If the advice on modification of duties has not been followed and in the absence of the individual's consent, the consultant in occupational medicine must inform the director of clinical services and the consultant responsible for infection control.

6.4.4 With the staff member's consent, detailed clinical information will be sought from his or her own physician. The consultant in occupational medicine will establish an ongoing relationship with the specialist to discuss modification of duties and co-ordinate care. This is particularly important if there are signs of AIDS-related disease, such as secondary infections and mental deterioration, which may prove hazardous in patient care.

6.5 Staff not involved in invasive procedures

The consultant in occupational medicine will discuss any alteration in work activity which may be necessary.

6.6 Ongoing supervision

HIV-infected staff who continue to work with patients must remain under close medical and occupational supervision. To this end, the consultant in occupational medicine will establish an ongoing relationship with the staff member's own physician to co-ordinate care. This is particularly important if the staff member is exhibiting signs of AIDS-related disease, such as secondary infections and mental deterioration, which may prove hazardous in patient care.

6.7 Confidentiality

The maximum possible level of confidentiality will be offered. In those cases where alteration of work is required, there will be the minimum necessary disclosure of information.

7. Employment issues

7.1 Recruitment, selection and training

The —— is committed to equal opportunities for all its employees. Applicants who have, or are suspected of having, AIDS/HIV should not be discriminated against with regard to recruitment, promotion, transfer or training. If they are deemed to be the most appropriate candidate for a post, the consultant in occupational medicine must assess their capability of carrying out the post on medical grounds and make appropriate recommendations regarding employment.

7.2 Employees with HIV infection/AIDS

7.2.1 Should a manager have cause for concern regarding an employee's health, the normal guidelines relating to sickness absence should apply and advice be sought from the personnel and occupational health departments.

7.2.2 Employees with HIV/AIDS who have problems carrying out the full range of their duties will be treated no differently from employees suffering from other illnesses whose health is affecting their work. It is important to note that the majority of individuals with HIV infection will be symptom free.

7.2.3 Where an individual's health is deteriorating to the point that they are unable to carry out the duties of their post, the usual considerations relating to cases of ill health will apply, e.g. retirement on grounds of ill health.

7.2.4 Where the occupational health department advises that an employee is capable of doing some form of alternative work, this possibility will be fully explored.

Section 2 Patient notification

I.   Introduction

1.1   These recommendations are based on guidance given by the Expert Advisory Group on AIDS (EAGA): *Practical Guidance on Notifying Patients*. This should be consulted for detailed procedures. Initial steps are outlined below.

2.   Confidentiality of health care workers

2.1   There is a general duty to preserve the confidentiality of medical information and records. Breach of the duty is damaging to the individual concerned, and his or her family, and it undermines public confidence in the pledges of confidentiality which are given to those who come forward for examination or treatment. In dealing with the media and in preparing press releases, it should be stressed that individuals who have been examined or treated in confidence are entitled to have that confidence respected.

2.2   There is, on the other hand, a duty to inform patients who may have been at risk of infection and take whatever steps may be necessary to provide reasonable reassurance. In the context of reassuring or treating such patients, it may be necessary to explain the circumstances which have given rise to concern. Legally, the identity of infected individuals may be disclosed with their consent, or wherever it is considered that patients need to be told for the purpose of treating their anxieties.

2.3   Such disclosure must be carefully weighed. EAGA considers that only in exceptional circumstances may disclosure without consent be justified. Those making such a disclosure may be required to justify their actions.

2.4   The fact that the infected health care worker may have died, or may already have been identified publicly, does not mean that duties of confidentiality are automatically at an end.

3.   Assess the situation

3.1   When the director of clinical services has been informed about an HIV positive health care worker, the following should be established:
   (a) employment history;
   (b) what, if any, types of invasive procedures are likely to have been performed;
   (c) how long the health care worker may have been infected.

3.2   Where there is doubt about the risk posed to patients, advice should be sought from the UK Advisory Panel on HIV-infected health care workers.

4.   Setting up an incident team

The district director of public health will normally lead the incident team. Members of the team will include:

   • Chief executive;
   • Consultant in communicable disease control;
   • AIDS services co-ordinator or counsellor;
   • Local consultant in microbiology;
   • Head professional from same speciality as health care worker;
   • Consultant in occupational health medicine;
   • Press officer.

5.   Notification of other bodies

The following should also be informed:

   • The district and/or regional chairman, as appropriate;
   • The regional director of public health;
   • The regional infectious disease epidemiologist;
   • The director of the appropriate Family Health Service Authority (as necessary);
   • The Department of Health – contact Dr Gwyneth Lewis or deputy on 0171–972 3355.

Chapter 6
Sickness absence

These are not sins of omission but signs of preoccupation.
(Michael Ondaatje)

Sickness absence, or absence attributed to sickness, is a major and increasing cost to organisations. A recent survey by the CBI estimated that over 160 million working days are lost through sickness each year. In this chapter details of the patterns of absence in different working groups and the different regions are given. The various methods of controlling absence, including sick pay schemes, the use of sickness absence statistics and recruitment checks, are discussed. An attempt is made to clarify the role of occupational medicine in the management of absence. Detailed information is given on the management of both long-term absence and intermittent persistent absence, with reference to employment and case law. The appropriate use of medical certificates and medical information is also discussed. The chapter concludes with a model policy for managing sickness absence.

Introduction

It is essential, at the outset of this chapter, that we are clear about what we mean by sickness absence. From the occupational health point of view, it is more realistically termed absence attributed to sickness. Although it is impossible to give any concrete information on the percentage of such absence which is genuinely the result of a medical condition, it is nevertheless clear that the frequency and length of any absence for an illness or injury are determined by many factors other than the actual disease process. For example, the individual's resilience and personality, the availability and suitability of treatment, domestic circumstances and the nature of the job all have a significant impact on the need for absence and its ultimate length. The employer obviously has little involvement in the medical management of the condition. The other aspects, however, can be considerably influenced by 'social' management.

Management commitment to controlling absence, which includes laid-down disciplinary procedures, a recognised sickness absence policy, good absence statistics and flexibility on rehabilitation and resettlement, has been shown to reduce sickness absence levels substantially (Audit Commission 1993). This is not only in the case of frequent short absences which, while possibly being related to genuine illness, may often be influenced considerably by the other factors already mentioned, but also in the case of long-term absence for recognised medical conditions, where active involvement of the employer in rehabilitation may produce results beneficial to all parties.

Demography

The most recent general review of absence from work was undertaken by the CBI and Percom. This review, entitled *Too Much Time Out*, was published in 1993 and was intended to be a follow-up of their 1987 review, *Absence from Work*. The more recent review covered 300 employers in the public and private sectors with a combined workforce of 1.22 million. It showed that in 1992 an average of eight working days per employee were lost owing to sickness. This represented 3.5 per cent of all available working time. It extrapolated to 166,712,000 total working days lost – far in excess of any other absence except annual leave. A cost to industry of at least £13 billion was estimated. Absence for manual workers in full-time work was almost twice that of non-manual workers, and absence levels in the public sector were on average 41 per cent higher than those in the private sector. See Table 6.1.

There does not seem to be any information on the reasons for higher absence in manual workers. It may be because of job requirements, employer attitudes or different sick pay arrangements.

However, there is a tremendous range of absence levels in the various sectors. For example, different local government rates ranged from 2 to 12 per cent. A review of sickness absence in London boroughs in 1991/92 (Audit Commission 1993) confirmed the variation, which ranged from 7

Table 6.1 Average number of days of sickness absence

| | 1987 | | 1992 | |
	Manufacturing	Services	Manufacturing	Services
Full-time manual	9	9	10	10
Full-time non-manual	5	6	5	6

Table 6.2 Percentage sickness absence by CBI region

Region	% sickness absence	No. of responses
Full-time manual		
Northern Ireland	5.9	12
Scotland	5.3	14
South Eastern	5.1	5
Northern	5.0	12
North West	5.0	16
Wales	4.8	20
Greater London	4.7	15
South West	4.6	8
Yorkshire and Humberside	4.5	8
West Midlands	4.3	14
Eastern	3.6	5
Southern	3.4	3
East Midlands	3.2	9
Full-time non-manual		
Northern Ireland	3.2	7
Scotland	3.0	28
South Eastern	3.0	19
Northern	2.9	12
North West	2.9	12
Wales	2.8	9
Greater London	2.8	3
South West	2.7	4
Yorkshire and Humberside	2.6	10
West Midlands	2.3	15
Eastern	2.3	12
Southern	2.3	14
East Midlands	1.7	9

to 17.4 days per employee. There are also regional variations, with the levels of absence being highest in Northern Ireland (5.9 per cent) and lowest in the East Midlands (3.2 per cent). See Table 6.2.

Significantly, smaller organisations have lower rates of absence. Such statistics confirm that the level of absence attributed to sickness is influenced by a number of factors unrelated to the disease process.

Controlling absence

Contract terms

The main legal considerations within the organisation are the terms of the contract of employment and related provisions such as sick pay. Obtaining medical information may be facilitated if the employee is contractually bound to 'submit to a medical examination at the request of the employer'. In certain job categories there are rigid rules of fitness and these should form part of the contract.

Sick pay schemes

Employers will normally require proof of incapacity for the provision of sick pay. Acceptance of self-certification, medical certification and certification from other parties is discussed later. With regard to entitlements, there are anomalies in certain companies where entitlement to sick pay may vary between that as a result of an accident at work and that because of general illness. It is necessary to point out here that the existence of a sick pay scheme should not be seen as confirmation of a specified period of absence to which the employee is entitled (*Coulson v. Felixstowe Dock and Railway Company* 1975). On the other hand, automatic dismissal at the end of the sick pay period may also not be seen as reasonable.

Disciplinary procedures

Clearly drawn-up disciplinary procedures are essential to prove that an employer has behaved reasonably. They also avoid the possibility of variations in treatment between departments and individuals. In the CBI survey 80 per cent of the organisations surveyed had clear disciplinary procedures.

Sickness absence policies

For the same reasons that disciplinary procedures should be clear, policies on the management of sickness absence are important. Such a policy facilitates management of absence cases, identifies the ground rules for employees and incorporates suitable management training. The policy must indicate at what point the manager should consider obtaining medical advice from the occupational health department, where one exists, or from external practitioners. It should also specify interventions, such as return to work interviews. In the CBI survey, 60 per cent of organisations had a sickness absence policy and 50 per cent a return to work interview. A model sickness absence policy is given at the end of this chapter (see pages 106–12).

Early retirement on the grounds of ill health

There is considerable variation in the hurdles which a sick employee has to surmount to be eligible for this type of dismissal. Entitlement related to length of service also varies considerably. Some superannuation schemes readily accept cases where the disability is peculiar to the requirement of the particular job, whereas others accept a more general disability. Incapacity continuing into the foreseeable future or permanent incapacity is essential.

Recruitment checks

A high percentage of organisations have some form of health and attendance check on recruitment. This may take the form of questions on illness and sickness absence in the general pre-employment questionnaire, or these questions may be contained within a medical questionnaire (confidential or otherwise). Some organisations still require pre-employment medical examinations or health checks. It is established beyond doubt that an employee's past record of attendance (particularly in relation to short-term frequent absences) is the single most important piece of information in determining the likelihood of regular attendance. Pre-employment medicals result in only about 1 per cent outright rejection. They are singularly unsuccessful in determining future absence levels and wasteful of resources in this respect. Of course, where there are significant health hazards or fitness requirements there are more compulsive reasons for pre-employment medical assessment.

Sickness absence statistics

Without comprehensive records and statistics, management of sickness absence is inevitably arbitrary and speculative. The CBI survey showed that 70 per cent of surveyed organisations kept computerised sickness absence records. Interestingly, sickness absence was 16 per cent higher in those companies which had only manual records. Sickness absence statistics need to be in a form which provides a tool for managers. The overall absence rate in their department gives them a baseline figure from which a target level can be drawn. However, to manage sickness absence different statistics are needed. The two most important are severity rate (the average number of days lost per person per year) and frequency rate (the average number of absences per person per year). Attention can then be focused on individuals with a frequency or severity rate above the department average. Most employers believe (rightly) that absence tends to be concentrated among certain individuals and certain groups. Group differences may be unavoidable because of different fitness requirements; for example, in health care workers or mine rescue teams. However, group differences can also be the result of custom and practice. An extreme example of this was in a group of overlookers in a cotton mill who took it in turns to be ill. This was particularly beneficial to all as, although the absentee was receiving sick pay, the others were given an equal share of his basic salary as they were covering his absence!

Occupational health

The role of occupational health in controlling sickness absence is often not understood by management. Occupational health physicians do not control or monitor sickness absence. This is the responsibility of the line manager. What occupational health should do is provide management with information on the existence of any underlying medical conditions, on the likely length of absence, on the likely frequency of absence, on the probable need for short-term modified work or long-term redeployment, and in general terms the likely effect of the medical condition on fitness for work. This is discussed in greater detail with regard to the management of sickness absence in the different categories of absence.

Disciplinary action and dismissal

Provided an employer acts reasonably throughout, employees may be fairly disciplined and dismissed for ill health or absence. The statutory test of reasonableness has been clearly defined (*Spencer v. Paragon Wallpapers Ltd* 1976). To be seen to be reasonable the employer must take into account:

- the nature of the illness;
- the likely length of absence;
- the need to have the employee's work done;
- the circumstances of the particular case.

Long-term illness needs to be considered differently from intermittent persistent short-term absences.

Long-term absence

It is perhaps important to restate here that the existence of a specified period of sick pay entitlement does not mean that action cannot be taken within this period. Similarly, the provision of a certificate of incapacity is only one of the factors that the employer must consider; it does not preclude dismissal. If absence of whatever length means that business cannot continue, dismissal can be held as fair.

Ill health is potentially a fair reason for dismissal if ill health affects the employee's ability to perform the work which he was employed to do. This refers to the job specification and not just to the work which he was actually doing at the time he became incapacitated. Likewise, if only part of the work cannot be carried out, the reasonableness of dismissing the employee will depend on how significant this part of the work is (*Shook v. London Borough of Ealing* 1986).

Obtaining medical advice

In order to come to a reasonable decision, it is obviously necessary for the employer to obtain medical information. This may be obtained from the occupational health physician, the individual's general practitioner or specialist, or an independent specialist. Where the doctor approached is responsible for the care of the employee, the provisions of the Access to Medical Reports Act 1988 apply. In practice this means that the employee must consent in writing to the report being requested, having been informed of his rights. He must be given the opportunity to see and agree the report and he has the right to request amendments. If the attending doctor does not agree to the amendments, the employee has the right to attach a personal statement.

When a report is requested the doctor should be told in writing the reason for the request and any possible outcomes. In the case of long-term absence it is usually essential to obtain information from the attending doctor. Where there is an occupational health unit, this information should be obtained through the unit, thus ensuring a fuller report while preserving medical confidentiality. Unfortunately, not all doctors are prepared to co-operate with employers in this respect and difficulty in obtaining their reports may result in considerable delay in resolving the case. It is sensible to state any fee offered for the report at the time it is requested to avoid embarrassingly large bills. The British Medical Association recommends a standard fee for such reports and the employer should not expect to pay more than this. Fortunately, many primary care doctors do not seek a fee in these cases as they appreciate the importance to their patient of contact with the employer.

Should the employee refuse consent for any medical information, he should be told in writing that a decision will be taken on the strength of the available evidence and that this could result in termination of employment. It is unusual for employees to persist in this refusal. Every effort should be made by the manager, the personnel department and the occupational health unit to reassure the employee of the fairness of any likely subsequent decision, and the need to determine the likelihood of future absence and long-term disability.

Employers are not bound to accept a medical report without question. A statement from a general practitioner that the individual is unfit to carry out the duties of the post, and should be retired on the grounds of permanent ill health, is unlikely to be accepted as sufficient evidence of long-term incapacity (*East Lindsey District Council v. Daubney* 1977).

Differences of opinion between the primary care doctor and occupational physicians are rare. When they do differ the two most common reasons are the general practitioner's lack of knowledge of the job requirements (he is usually relying on the employee's job description) or the general practitioner's desire to help and please his patient, who may well feel that he no longer wishes to continue in employment and is seeking a beneficial mode of exit. Employers are not required and not competent to decide between medical opinions. They are entitled to accept the opinion of their own medical advisers (*Ford Motor Company v. Nawaz* 1987). Where some doubt remains the opinion of an independent consultant should be obtained.

If medical advice indicates that some less arduous task will be required in future, the employer must consider alternative employment. The employer is, however, not expected to go to unreasonable lengths to accommodate the employee.

Consultation

In cases of long-term sickness, warnings regarding attendance rates are clearly inappropriate and may, in fact, be detrimental to the employee's well-being. Having obtained medical advice, it is reasonable for the employer to discuss the position with the employee, so that he is given a chance to state his case. Only in rare circumstances would consultation be unnecessary, such as when contractual fitness standards can no longer be achieved. Although in practice it is unlikely to affect the decision in such a case, it is nevertheless good management practice to meet and discuss the reasons for dismissal.

Intermittent persistent absence

These cases are usually much more difficult to define and manage. If, as is usual, the reasons for absence are all minor, unconnected ailments, there is no legal requirement to obtain medical advice. However, this sort of absence may mask an underlying significant health problem or work-related disorder. It is therefore sensible for the manager to seek medical advice. This is more readily available where there is an occupational health service. In such cases the employer needs to consider:

- the nature of the illness;
- the likelihood of this illness recurring or of some other illness;
- the length of the various absences and the spaces of good health and performance in between;
- the need for the work to be done;
- the impact of the absences on other employees;
- the extent to which the employee has been made aware of acceptable standards.

The employee may indicate that there is an underlying health problem. No useful purpose may be served by obtaining medical evidence except in these cases, as there is no underlying medical condition, and it is impossible to verify the situation after the employee has returned to work.

Sickness absence statistics are particularly useful in managing cases of frequent absence because they provide management with acceptable standards of attendance.

Employees should be interviewed and made aware of these standards. They should be warned that, if they do not meet these standards, dismissal may result (*International Sports Company Ltd v. Thomson* 1980). They should then be given a date when their record will be reviewed. During this period it is important to maintain proper attendance records. If at the end of the review period there has been no substantial improvement, the employer is entitled to dismiss the employee. This dismissal will not be on medical grounds but on the grounds of unacceptable attendance. Of course, the situation is rarely so clear cut; employees often achieve a better record during the monitoring period but subsequently revert to the previous high level of absence.

Where, on interview, the employee indicates that there is an underlying health problem, medical information should be obtained as in cases of long-term absence. Consideration should be given to:

- the relationship of the underlying health problem to the absences;
- the likely resolution of the underlying problem;

- the prospects of normal attendance;
- the appropriateness of early retirement on medical grounds.

Malingering

Malingering is a word which is not favoured by the medical profession. There are undoubtedly individuals who are prepared to 'play the system'. However, they are exceptional. Most cases of frequent absence without any real sickness component are the result of genuine social problems. These may be concerned with the care of children or elderly parents. Sickness absence is generally more 'acceptable' than absence for humanitarian reasons. It is also paid whereas social absence is not. The occupational physician regularly finds that individuals with frequent sickness absences are using these to cover unresolvable domestic pressures. General practitioners will often certificate such absences with a diagnosis such as 'anxiety state'. In these circumstances this is not far from the truth. The occupational health physician must judge whether the underlying situation will soon be under control, and the likelihood of future satisfactory attendance. Managers may be uncomfortable with this resolution of the problem, but some flexibility can often result in the return to full attendance and competence of a valued employee.

Use of medical certificates

It is up to the employer whether or not to accept that an employee is incapable of work. Most companies require some form of certification before paying occupational sick pay and Statutory Sick Pay. General practitioners are not required to issue a certificate during the first seven days of absence. The employee may produce a certificate from someone other than a medical practitioner; for example, an osteopath, a physiotherapist or a psychologist. Such certificates must be considered on their own merits. The employer is entitled to request a certificate from a medical practitioner, but certificates from other professionals are usually accepted by the DSS.

As general practitioners are not required to issue a certificate during the first seven days of absence, some form of self-certification is necessary. Most organisations have their own self-certificate and these should be completed by the employee on return to work. If this is accompanied by a return to work interview, managers may be able to assess the reasons for absence more realistically and facilitate control measures.

Medical certificates should be regarded as strong evidence of incapacity, but they give the employer little information. Most absence is, in fact, self-certificated. How can the doctor do anything but accept the patient's description of, for example, back pain or diarrhoea? It is difficult for a doctor to refuse a certificate when the patient is adamant that he has such symptoms. A similar problem arises with long-term absence. There will probably be objective evidence of incapacity such as a surgical intervention or a well-documented medical condition. But the decision as to when the patient is fit to return to work is largely the patient's own. An enthusiastic employee, eager to return to work, can give the doctor a biased description of the 'lightness of his duties'. The unenthusiastic worker can, on the other hand, build up a burdensome job description. Where there is an occupational health unit, it is worth while involving this unit at an early stage in long-term absence. This will provide a realistic appraisal of the employee's capacity to undertake the work. The occupational health practitioner has the advantage of understanding both the disease and the job requirements. In general, managers tend to give too much weight to the medical certificate. It may be issued for reasons other than medical incapacity which have already been described:

- The doctor may not understand the job.
- He or she may feel that the patient's wishes are paramount.
- He or she may wish to maintain a good relationship with the patient.
- He or she may not be concerned about the veracity of the certificate.

Conclusion

Failing to manage sickness absence in one's employees is not humanitarian or paternalistic – indeed, quite the reverse. If employees feel that it does not matter whether or not they appear for work, their sense of their own worth and the worth of their jobs is considerably reduced. If a job is so unpleasant or boring that the employee is not prepared to get out of bed for it, the solution is not to be found in allowing uncontrolled absence, but in improving the job. Employees who give no priority to attendance at work must be shown that their attendance is important, and that their absence is noted and unacceptable. Where there is genuine illness, concern and consideration should be shown by a return to work interview and by maintaining contact during absence. Employees should be given every assistance to attain an acceptable attendance record, but they must realise that there is no room for persistent non-attenders in the tight budgetary provisions of today's workplace.

A policy for managing sickness absence

1.   Introduction

1.1   The purpose of this policy is to give managers the knowledge and confidence to deal with sickness absence. It should be handled in a positive way and management must be seen to be supportive of their staff.

1.2   The procedure is by definition broad because every case of sickness absence will be slightly different from others. However, the overriding test will be: 'Was management action reasonable in the circumstances?' In view of the above, personnel advice should always be sought at an early stage.

1.3   In most cases sickness absence will be genuine and needs to be dealt with sensitively. Occasionally, individual employees may take advantage of the sick pay scheme and this clearly affects the running of the department and staff morale adversely.

2.   Staff records

2.1   Staff records must be kept by the head of department relating to the length of, and reason for, sickness absence. It is important that such details are recorded, even where absence is for a single day or uncertified. Managers should ensure that their sickness absence reporting procedures allow for such details to be ascertained. The pattern of sickness is also relevant. Monitoring should be regularly undertaken by heads of department in conjunction with the personnel department.

2.2   Absence levels – action required

As a general guide, absence on six or more occasions in a 12-month period should be acted upon. Long-term absence of four continuous weeks should also be acted upon.

Managers will be assisted in working out average sickness rates for their own department, which can then be used in discussions with staff regarding sickness absence.

2.3   Referral to the occupational health department

Any staff referral to the occupational health department should be by completion of the standard form for this purpose, together with the sickness record and job description (where appropriate).

3.   General principles

3.1   Prevention

Whenever possible, sickness absence problems should be prevented from occurring by:

3.1.1 Identifying any health risks of the job and possible aids to prevention (e.g. heavy work may induce back problems).

3.1.2 Discussing with applicants, at interview, their patterns of absence in previous employment. Health questionnaires should be reviewed by the occupational health department prior to the appointment being made. Previous employment references may be used to verify the past absence record.

3.1.3 Counselling employees early when an unsatisfactory absence record is developing.

3.1.4 Meeting briefly with any employee returning to work after sickness absence (even single days). The meeting should not be threatening, but will show the employee that his or her attendance is valued (and, therefore, that he or she is missed when absent). It may also provide an opportunity for the employee to highlight difficulties or problems at work or home, or underlying health problems. Normally, the supervisor or immediate manager should be responsible for holding these meetings.

3.2   Frequent short-term absence and misconduct

3.2.1 When staff take frequent short-term absences attributed to sickness, they may be genuinely ill but there is a possibility that factors other than health are influencing their level of absence. This may constitute misconduct.

3.2.2 When dealing with 'misconduct' you must refer to the disciplinary procedure. You will need to consider the possibility of referral/advice from the occupational health department. Normally, you should ensure that the employee has been assessed by the occupational health department as having no detectable underlying medical problems causing absence *before* taking any normal disciplinary action.

4.   Do not delay in taking appropriate action

4.1   In order to ensure the effective running of your department it is important to take timely appropriate action. This will also ensure consistency within your department and between departments.

4.2   Managers may refer staff to the occupational health department, discuss the matter with that department, or take other appropriate action, even when staff produce regular medical certificates. Referrals should be made in writing, using the standard form, giving details of dates of sickness absence and reasons, if known. Where managers are unsure about the appropriateness of referral, they should first telephone or write to the occupational health department. In addressing problems of sickness absence, managers need to distinguish between frequent short-term absence and prolonged sickness absence, as set out below.

5.   Categories of sickness absence

5.1   There are two main categories of sickness absence:

- short-term frequent sickness absence;
- prolonged sickness absence.

5.2   Short-term frequent sickness absence

5.2.1 The manager should discuss the problem with the employee at the earliest opportunity to establish whether:

(i)   there is any underlying reason for the absence;
(ii) improvement is imminent;
(iii) any treatment is being received;
(iv) there is any employment-based contributory cause;
(v)   there are any alternatives which may improve the situation, e.g. changing the hours of work.

5.2.2 Where appropriate, after discussion with a personnel officer, the employee should be referred to the occupational health department. Full details of the employee, his or her absences and any known reason for absence should be given.

5.2.3 If, at the end of the monitoring period, there is no significant improvement, there should be a further meeting between the manager and the employee to review the points listed in 5.2.1 (i–v), and to determine whether there is a new health problem. If the latter is the case, the employee should be referred again to the occupational health department. If there is no valid reason for the absence, the appropriate stages of the disciplinary procedure should be followed.

5.2.4 If the occupational health department advises that there is no detectable underlying problem, the employee should be counselled further, informed of the standards expected, and advised that the disciplinary procedure may be invoked. The employee should be told that continuation of this level of absence could ultimately result in termination of employment (on the grounds of incapability or misconduct). Following a warning, the employee's sickness record should be monitored carefully and review dates arranged, as appropriate.

5.2.5 If the occupational health department advises that there is a detectable problem, the manager will be advised in general terms how the ill health is to be resolved.

5.2.6 The length of monitoring periods will depend on individual circumstances and the post held by the employee.

5.3 Prolonged sickness absence

5.3.1 In these cases, the employee should be referred to the occupational health department (normally after four weeks) for assessment on the nature of the sickness and whether any improvement is likely. At this stage, it may not be necessary for the employee to be seen by the occupational health physician. If the manager needs advice as to the best course of action, he may complete the referral form or first contact the occupational health department. It may be necessary to obtain information from the treating practitioner. The employee will be asked to sign his or her agreement that information from his or her GP or specialist may be sought by the occupational health department. The employee will also be informed of his or her rights under the Access to Medical Records Act 1988.

5.3.2 Action following a medical report

The occupational health department will indicate any of the following:

(i)   The employee is now able to return to work.
(ii) The employee will be able to resume work in a reasonable period of time.
(iii) No improvement is likely.
(iv) The employee will be able to return to work subject to certain conditions.
(v)   It is too early for the occupational health department to give a definitive view and a review of the employee's health will be undertaken within a specified time period.
(vi) The employee should be redeployed either temporarily or permanently (see Section 7), or given long-term alterations of duties in the present post.

5.3.3 Depending on the medical report, the following options should be considered:

(i) If the occupational health physician believes that there is no significant health problem, the employee should be advised that he or she should return to work to carry out his or her contract of employment satisfactorily. Failure to do this will ultimately result in termination of employment.

(ii) Rehabilitation into the job, e.g. working part time initially or undertaking restricted duties for a short period of up to four weeks. (See Section 6.)

(iii) Redeployment to an alternative post. (If the post is of a lower grade, protection does not apply: see Section 7.)

(iv) Early retirement on the grounds of ill health. (See Section 8.)

(v) Dismissal on grounds of incapability. (See Section 9.)

5.3.4 If the problem continues, it may be appropriate to refer the employee again to the occupational health department, using the referral form. The employer retains the right to obtain a further medical opinion from a relevant specialist which will be arranged through the occupational health department. The occupational health department provides advice regarding future employment, knowing the sickness problem of the employee and details of the work. If the employee refuses, for any reason, to undergo an examination, either by the occupational health department or another relevant specialist, this could be construed as misconduct because of a contractual commitment to undergo a medical examination. Any decisions taken regarding the employee's future employment would be without the benefit of such information and the employee should be advised accordingly. In certain cases, a report from the treating doctor may be adequate. The view of the occupational health physician should be sought before taking further action.

6.   Rehabilitation

6.1   This is designed to enable the employee, following long-term sickness absence, to undertake gradually the full range of his or her duties. This may include carrying out alternative duties or working reduced hours, but normally only for a limited period of up to four weeks.

6.2   The terms should be discussed with the personnel department.

7.   Alternative work

7.1   Advice should be sought from the occupational health department on the decision as to whether to suggest to the employee that, although he will not be fit to carry out the duties of his post, an alternative job may be appropriate. The occupational health department should advise on suitable posts. The alternative work should be discussed with the employee and arrangements made for him to be interviewed by the appropriate head of department. Wherever possible, the employee should be given favourable consideration. If there are no suitable posts within the organisation, the personnel officer should, with the employee's agreement, write to other branches which are convenient for travelling to the employee and establish whether they have any suitable posts.

8. Ill health retirement

8.1 Having received confirmation from the occupational health department that the employee is not likely to return to work, is unfit and expected to remain unfit to fulfil the duties of his post, the head of department/ manager and personnel officer should meet with the employee to explain the situation and suggest early retirement (if the employee has two years' superannuated service). An application for ill health retirement will then be made by the personnel officer.

8.2 If the ill health retirement application is approved, a representative of the finance department will meet with the employee and complete the retirement application form.

8.3 On termination of employment, the employee will receive a lump sum pension payment and regular pension payments at specified intervals. As the employee has applied for ill health retirement, a notice period is not applicable.

8.4 If the application for ill health retirement is rejected, the procedure for dismissal on grounds of incapability may be followed.

9. Dismissal on the grounds of incapability

9.1 If early retirement on the grounds of ill health is not possible, or considered inappropriate, the personnel officer and the manager should meet together to consider dismissal of the employee on grounds of incapability, having first given consideration to continuing to employ on alternative employment or changed hours, if appropriate and possible. (Note that only dismissing officers designated under the disciplinary procedure may terminate an employee's contract of employment.)

9.2 An employee dismissed for incapability has the right of appeal against the decision of the organisation. Those employees with at least 104 weeks' continuous service can also apply to an Industrial Tribunal, claiming unfair dismissal. The Tribunal will consider whether or not, in the circumstances, the employer acted reasonably in treating incapability as a sufficient reason for dismissing the employee. The Tribunal will expect that management has taken the following facts into account:

(i)   the nature of the illness;
(ii) the employee's length of service;
(iii) the expected length of a continuing absence;
(iv) the effect of absence on the department's operational efficiency;
(v)   the sickness record.

10. Advice

Advice should always be sought from the personnel department to ensure consistency of approach within current legal regulations.

Chapter 7
Smoking in the workplace

How can the smoker and the non-smoker be equally free in the same railway car?

(George Bernard Shaw)

It has been estimated that 50 million working days are lost to British industry every year from smoking-related sickness absence. The increasing evidence that passive smoking may also lead to illness and death means that tobacco smoke could now be regarded as a toxic hazard. The responsible employer should initiate a programme to minimise smoking in the workplace, not only for the promotion of health but also in line with the requirements of health and safety legislation. This chapter looks at the current legal position and explores ways of producing a smoke-free atmosphere in the workplace. Rest rooms and rest areas must include suitable arrangements to protect non-smokers from discomfort caused by tobacco smoke. Guidance on the development of a formal written smoking policy is given and the case for encouraging all employees to increase their health awareness is outlined.

Introduction

There can be little doubt that smoking in the workplace is an issue with which management must engage. It has been estimated that 50 million working days are lost to British industry every year from smoking-related sickness absence. For this reason alone the responsible employer should look for initiatives which will discourage smoking in the workplace.

The use of tobacco is ubiquitous. About 13 million adults in the UK smoke cigarettes, accounting for 80 per cent of tobacco consumption.

There are two types of tobacco smoke: mainstream smoke, inhaled and exhaled by the smoker, and sidestream smoke which is released directly into the atmosphere from the cigarette. This contains a higher concentration of the tobacco chemicals, including known carcinogens. Sidestream smoke forms the greater part of environmental tobacco smoke (ETS). Lung cancer, which has been linked to smoking, kills over 40,000 people in the UK each year, 90 per cent of whom are smokers. Other forms of cancer are also more common in smokers. However, cardiovascular disease, including heart attacks, remains the major cause of death in smokers.

Other health effects of ETS are recognised including irritation and discomfort of the mucous membranes, exacerbation of adult asthma and increased incidence of heart disease.

Legal considerations

Since ETS is now recognised as a health hazard the requirements of health and safety legislation apply. This has not yet been tested in the courts. In 1993 there was an out-of-court settlement between Veronica Bland and Stockport Borough Council in which Mrs Bland claimed that her health had been damaged by ETS. It is known that further cases are in preparation. Relevant legislation includes the following.

Health and Safety at Work etc. Act 1974

This Act underpins much of the other health and safety legislation. It requires all employers to promote and ensure the health and safety of their employees and provide a safe workplace.

Workplace (Health, Safety and Welfare) Regulations 1992

These are part of a group of six regulations (the six pack) promulgated in 1992 and stemming from the EC Framework Directive. Regulation 25(3) states: 'Rest rooms and rest areas shall include suitable arrangements to protect non-smokers from discomfort caused by tobacco smoke.'

Existing current work areas must comply by 1 January 1996, while new or significantly altered workplaces must comply as soon as they come into use.

Management of Health and Safety at Work Regulations 1992

Another one of the 'six pack', these impose a duty to assess all risks to the health and safety of the individual worker and eliminate the hazards present. There is no doubt that the risk associated with the hazards presented by passive smoking must be assessed and reduced.

Providing an atmosphere free from tobacco smoke

Apart from the obligation to produce a smoke-free atmosphere, the employer may see other, more general, advantages in providing such an atmosphere, namely a reduced fire risk and reduced absenteeism. On the negative side, companies should be aware that implementing a smoking policy will cost both time and money: time in consultation, negotiation and attendance of counselling/support groups; money for the purchase and dissemination of posters and educational literature. The only certain way of achieving a smoke-free atmosphere is to ban smoking in the workplace. It is not recommended that this be imposed by a top-down edict. On the contrary, it is strongly recommended that the workforce are made familiar with the legal requirements and the health case so that they can take part in the decision-making, implementation, monitoring and development of the smoking policy in an informed way. It is argued that to implement such a change, without detailing the provisions and facilities for smokers, including those designed to assist smokers who wish to stop, would be seen as unfair, and would call for a high level of subsequent workplace supervision.

Development of a smoking policy

Action on Smoking and Health (ASH), in collaboration with other interested parties including the Health Education Authority, has produced a guide to good practice in workplace smoking. The steps recommended in the development of a company smoking policy are:

- preparing for action;
- taking action;
- launching the policy;
- making it work.

Preparing for action

The best way of ensuring that action on smoking at work is successful is to make the smoking policy reflect the wishes of the workforce, as far as possible. One of the best ways of doing this is to circulate a questionnaire (the purpose of which is explained in advance) to the whole workforce or a representative sample.

One of the most effective ways of organising a questionnaire and planning subsequent action is to set up a small working party. Such a party should represent the views of smokers and non-smokers, and should include representatives of management and, where recognised, trade unions.

Before any action is taken, it is important to review current practice by asking the following questions:

- Is there any existing policy on smoking, written or otherwise?
- In which areas is smoking currently restricted and why?
- Do some groups of workers seem to smoke more heavily than others? Does this relate to working conditions?
- Have there been any cases of:
  - non-smokers complaining of a smoky atmosphere or bad ventilation?
  - smokers unduly restricted in areas where they can smoke?
  - smokers asking for assistance in giving up?
- Who did these workers complain to, and what happened?
- Is there any special hazard?

The nature of the industry or work may make smoking particularly dangerous: for example, the risk of contracting lung cancer in a smoker who is also exposed to asbestos dust is around 55 times greater than that of an ordinary non-smoker, since the effects of asbestos and tobacco smoke chemicals appear to be cumulative. Specific factors such as these may be strong motivators.

The answers to these questions, in conjunction with the questionnaire, help to indicate the kind of policy which may be suitable. It is also important, however, to clarify why action is being taken and its aims.

Questionnaires

The questionnaire requires a great deal of thought as it must avoid any suggestion of bias in the way questions are phrased and be seen as appropriate and relevant to the people to whom it is addressed.

Each organisation will need to develop a questionnaire appropriate to its own requirements. A letter explaining the purpose of the questionnaire helps to establish the reason for its circulation and the possible outcomes. A questionnaire should also attempt to establish prevailing staff attitudes. Staff must also be assured that the results of the survey will be fed back to them. A sample questionnaire and other written communications to employees are given at the end of this chapter (see pages 119–22). The working party should set a timetable for assessment, the reporting of results, proposals for a smoking policy and implementation.

Taking action

Written policies

The advantages of a formal written policy are self-evident: it is a reference document, an indication of commitment, a gesture of encouragement and a definition of responsibility. (A sample policy is given at the end of this chapter: see page 123.)

Depending on the internal administrative practices of the organisation, a policy statement may be supported by a company manual providing background information, advice, administrative instructions, responsibilities and accountabilities. An extract from a management manual is given at the end of this chapter (see page 124).

According to the National and Local Government Officers' Association (as it then was), there are six clear negotiating options for restricting smoking in working areas:

- *Opt-in scheme*: smoking is allowed until one or more persons object.
- *Each office/department/section* is to determine its own restrictions.
- *Opt-out scheme*: unless everyone is in agreement that it should be allowed, smoking is banned in all working areas.
- *Majority decision*: the working area is designated smoking or non-smoking by a majority vote.
- *Smoking is restricted to specific times*: for example, during tea or coffee breaks.
- *Total ban*: smoking is prohibited at all times in all working areas.

Once a policy has been established, it is clearly important to encourage compliance with it. The disciplinary procedure for those who break the restrictions should be well publicised and the success of the policy should be monitored regularly. Once established, a smoking policy should not be deemed to be written in tablets of stone: monitoring its effectiveness must involve considering whether or not it needs to be changed in the light of changing smoking patterns, or in response to any problems which have emerged.

Launching the policy

The success of a smoking policy can be helped by careful timing of its launch and by taking advantage of internal and external publicity. For example, companies may consider it advantageous to co-ordinate the launch with National No-Smoking Day or with the Budget – both occasions which generate considerable media and public interest in smoking. Outside agencies can help; ASH and the Health Education Authority both produce useful literature – either free or at a low cost – which can be circulated to employees. The launch should be well publicised internally by the prominent display of notices. An exhibition could be mounted, and participatory exercises – such as lung function tests – could be made available. Companies could consider the advantages of introducing a smoking policy as part of a general campaign towards increased health awareness. Integrating a smoking policy in this way perhaps prevents smokers from feeling 'victimised' and encourages all employees to take a greater interest in health issues in general.

Making it work

The old adage that 'the care and effort put into the management of the change will reflect in the quality of the implementation' applies. Commitment to the new circumstances should be as obvious as possible. The policy will need to be supported in all parts of the organisation. This commitment will require a consistent approach and a willingness to invest time and money, and to subordinate the short-term interest for the longer-term good. It is recommended that, if a working party has been used in the research, development and implementation of the policy, it should continue to meet and have responsibilities until such time as the new situation has become a way of life and can safely be left to general consultative, participative arrangements. One way of showing commitment is by regular follow-up to check staff satisfaction and developments.

Of particular benefit is National No-Smoking Day which will offer not only resources for an in-house programme but provide aims and objectives, ideas for action, and a range of printed images and factsheets which can be used in a local campaign, as well as posters, leaflets and stickers.

Conclusion

Like good law, the achievement and maintenance of a responsible smoking policy, and good practice within an organisation, will benefit from the belief that the policy is necessary, fair, balanced and appropriate.

Experience shows that the perception of these factors relies on a well-considered strategic approach: one which involves and takes notice of the views of those to whom the policy applies, non-smoker and smoker alike.

Introductory letter

Dear Colleague

Developing a smoking policy

You may have seen the recent publicity in the press and TV about the possible hazards to a person's health from exposure to the smoke of an active smoker – 'passive smoking'.

Employees have raised the question of this health risk through the Health and Safety Committee. The Committee agreed that a working party of employees should look into the problem, and, in the interests of employees' health and safety, recommend whether the company should introduce a policy which (a) restricts smoking at work in some form and (b) provides support to people who want to stop smoking.

The members of the working party are:

Dr ——, the company doctor, has produced a brief summary on passive smoking and the current view of the health risks involved. This is attached for you to read.

Before considering any form of policy on smoking, the working party wants to know the views and preferences of employees across the company. We would be grateful, therefore, if you would complete and return the attached questionnaire.

This could affect *you*, so please take a few minutes to complete and return this form. It is important that we know what *you* think about smoking at work.

Thank you in anticipation of your co-operation.

Yours sincerely

For and on behalf of the Smoking Working Party

Questionnaire on smoking at work

Please complete this questionnaire and return it in confidence to the personnel department.

---

1. Should there be more restrictions on people smoking at work than there are at present?

   A. Yes.
   B. No.

---

2. Which best describes your view of what the company policy on smoking should be?

   A. A total ban on smoking on site at all times.
   B. A total ban on smoking on site except in designated 'smoking areas'.
   C. Smoking permitted with increased restrictions in work and rest areas.
   D. Each work area should be allowed to make its own decision whether or not to restrict smoking, i.e. a majority vote.
   E. No change to the current situation.
   F. Other; please specify.

---

3. Which of these best describes your view about smoking in *your* work area?

   A. There should be no smoking except at break times.
   B. Smoking should not be allowed.
   C. There should be separate areas where smoking is permitted.
   D. Smoking should be allowed in all areas.
   E. There should be a majority vote on whether smoking is allowed.
   F. Other.

---

4. For each of the listed areas, specify your preferred choice. The choices are:

   A. No restrictions on smoking.
   B. Separate areas for smokers.
   C. Smoking permissible only at specified times.
   D. A total ban on smoking.

The areas are:

Reception area	A	B	C	D
Corridors	A	B	C	D
Meeting rooms	A	B	C	D
Canteen	A	B	C	D
Rest rooms	A	B	C	D
Toilets	A	B	C	D
Company vehicles	A	B	C	D

---

5. Which of the following describes your working area best?

   A. Private office.
   B. Private office but used by others outside your working hours (i.e. shifts).
   C. Office shared with small number of other people.
   D. Open plan office.
   E. Shop floor.
   F. Other; please specify.

---

6. Which of the following describes you best?

   A. I am a smoker who wants to give up.
   B. I am a smoker who doesn't want to give up.
   C. I am an ex-smoker.
   D. I am a non-smoker.

To: All employees

From: The Smoking Working Party   1 December 199X

Summary of results of questionnaire on smoking

A total of 387 people replied: 69.5 per cent indicated that there should be more restrictions on smoking at work; 28.9 per cent indicated that there should be no change.

The individual views of respondents on a company smoking policy were as follows:

(a) A total ban on site	20 per cent
(b) A total ban except in designated areas	36 per cent
(c) Permitted smoking with increased restrictions	17 per cent
(d) Majority vote for each work area	9 per cent
(e) No change	18 per cent

Of the respondents, 27 per cent were smokers, half of whom want to give up, and 67 per cent would like help to give up.

Seventy-nine per cent of respondents indicated concern over health problems associated with smoking. Thirteen per cent felt the need to move away frequently from smokers at work and 37 per cent only occasionally felt the need to move away.

Sixty-five per cent of smokers who responded felt that it would be difficult if they could not smoke during working hours.

Sixty-eight per cent of respondents believe that the question of smoking should not be considered when recruiting new employees.

The Committee will now undertake an in-depth analysis of the questionnaire and draft a policy on smoking at work for discussion at the Consultative Health and Safety Committee.

A smoking policy

We would like to draw your attention to the fact that this company has a smoking policy which represents the views and wishes of our employees.

The main aim of the policy is to ensure a smoke-free atmosphere for employees and visitors.

The policy restricts smoking to specific designated areas throughout the company and these are clearly marked.

If you wish to smoke, you are therefore requested to smoke only in one of these areas.

Thank you for your co-operation.

A manager's manual on smoking policy

Introduction

The company has adopted a formal policy to reduce workplace smoking; this complements existing health and safety commitments to provide a healthy environment and promote personal health care within the company.

Policy

Cigarette smoking is the most preventable cause of premature death and disability in the UK. Therefore, in consideration of the health and well-being of all its employees, the company is committed to the development of programmes to reduce smoking in the workplace and to assist smokers in their efforts to stop smoking.

Responsibility

Disciplinary action

Any person smoking in a non-smoking area will be counselled within the scope of the company disciplinary procedure.

Annual review

The policy will be reviewed annually.

Why are we implementing this policy?

We aim to safeguard and improve the health of our employees by:

- promoting awareness of health hazards;
- protecting non-smokers from smoke by reducing the effects of passive sidestream smoke;
- encouraging smokers to stop.

General

This policy is intended to increase all employees' awareness of the dangers of smoking and maintain employee health throughout employment with the company. Employees and prospective employees should be made aware of this policy with the specific aim of seeking their support, and it is their responsibility to ensure it is maintained.

Chapter 8
Health promotion in the workplace

Accept change or remain behind in a state of shock.

(Alvin Toffler)

The Health Of the Nation targets five areas: coronary heart disease, cancers, mental illness, HIV/AIDS and accidents. All these are important and conducive to education in the working age group; initiatives in the workplace are seen as part of the strategy. This chapter looks at the business and the moral case for health promotion in the workplace. The cost to industry of unhealthy lifestyles is explored and some attempt is made to demonstrate the effectiveness of health education. The chapter goes on to consider the appropriateness of health promotion activities, including screening, education and the creation of a healthy environment. The criteria for workplace screening are outlined, with reference to the government's target areas. The development of an action plan is described, including assessing organisational needs and defining relevant target groups.

Introduction

There are about 26 million people in full- or part-time work in the UK, spending up to 60 per cent of their waking lives in their places of work. The government publication *The Health of the Nation* has targeted five areas where the nation's health can and should be improved, to bring it into line with comparable countries. These are:

- coronary heart disease and stress;
- cancers;
- mental illness;
- HIV/AIDS and sexual health;
- accidents.

None of the targets proposed can be achieved by action solely at work, although all are important to the working population. Of course, employers cannot take responsibility for the lifestyle of their employees, although they can reasonably be expected to aim for no work-related ill health or accidents.

In the same document the Department of Health has indicated that it will be exploring ways of creating national and local networks to draw together scarce public health skills. For smaller organisations these sorts of initiative may be the only cost-effective way of providing health promotion for their employees.

The business case for health promotion

At shop floor level the increased use of information technology and closed systems of work have tended to increase the value of individual members of the workforce. The complexities of organisations and their speed of change, which characterise the 1990s, require good management with well-developed special skills. Organisations, therefore, cannot afford to lose good personnel in whom much money may have been invested. Nor is it possible for companies to function efficiently if less than optimum performance results from deleterious lifestyle and poor health.

Information on the benefits of health promotion in the workplace is sparse in relation to the enormous amount of activity in this area. Many studies have looked at the impact of drinking and smoking in the workplace. Ryan *et al.* (1992), in a prospective study of 2500 postal employees, calculated the relative risk between smokers and non-smokers in four areas: turnover, accidents, injuries and indiscipline. See Table 8.1.

Ryan *et al.* showed that smokers had 25 per cent more accidents and when these occurred they had 40 per cent more chance of injury. Examples of indiscipline increased over 50 per cent. There appeared to be no related increase in turnover. They also demonstrated an increased sickness absence level in smokers, the mean rate being 5.43 per cent for smokers compared with 4.06 per cent for non-smokers. Similarly, Gabel (Gabel and Colley Niemeyer 1990), looking at staff in a public

Table 8.1 Increased employment risk associated with smoking

Relative risk	Smokers : Non-smokers
Turnover	1:01
Accidents	1:29
Injuries	1:40
Indiscipline	1:55

health agency, showed that smokers had more sickness absence, took less exercise and had a poorer diet.

Jenkins undertook a six-year longitudinal study of the occupational consequences in white-collar workers of drinking over the safe limit of alcohol (Jenkins *et al.* 1992). He showed that even moderate alcohol consumption substantially increased sickness absence and was associated with lack of promotion for men. In a detailed study of the Du Pont company, Bertera (1991) demonstrated an excess annual illness cost per person associated with five behavioural risks. See Table 8.2.

Table 8.2 Behavioural risks

Excess annual illness cost per person	$
Smoking	960
Obesity	401
Excess alcohol	389
Raised cholesterol	370
Raised blood pressure	343

The only behavioural risk not associated with a significant increase in sickness absence costs was lack of exercise. The Royal College of Physicians has estimated that 50 million working days are lost each year from smoking-related diseases. The likely cost in terms of lost production is over £2500 million. Alcohol-related disease is estimated to cost £700 million in sickness absence alone.

The case for the cost to industry of poor lifestyle has been made. What is less convincing is the effect of health promotion and the possible benefit to individual organisations. Generally, outcomes have not been measured, although Leviton (1989) has developed a model to assess the likely benefits of health promotion in a number of areas. He showed that screening for high blood pressure paid for itself. In some organisations smoking cessation programmes were also self-financing. However, he was unable to demonstrate any significant payback from cholesterol screening. The Johnson & Johnson and Du Pont corporations have both been able to demonstrate a reduction in sickness absence costs (between 9 and 20 per cent) following the introduction of an employee assistance programme.

There are probably a number of reasons why the effect of health promotion in the workplace has been poorly measured. The workplace does not lend itself to randomised studies, hence the lack of controlled ones. It is inevitable that any health promotion activity aimed at volunteers in a general workforce will receive more attention from those who are already aware. The intervention group is, therefore, considerably biased. Staff turnover interferes with long-term follow-up and organisational change may have an independent impact on outcomes. Few organisations have good epidemiologically based sickness absence statistics. Finally, there may be a considerable time-lag before health effects can be identified.

The outcomes that might be expected from successful health promotion are:

- a reduction in sickness absence;
- reduced staff turnover;
- improved productivity;
- improved morale;
- a reduction in the loss of key personnel;
- enhanced recruitment.

Other reasons for introducing health promotion are: specific fitness requirements for certain jobs, such as fire brigade personnel and catering staff; as an adjunct to statutory requirements new health and safety legislation is tending more towards individual fitness for the job. For example, the Manual Handling Operations Regulations require an assessment of the operator's fitness for each task.

All the outcomes would, of course, be important gains, but it is likely that organisations which embark on a full health promotion programme may have other motives which one could describe as the moral case.

The moral case for health promotion

Since the business case for health promotion has not yet been proved, why should organisations promote health at work? There are a number of reasons which could be described as purely altruistic and which demonstrate a more holistic approach to workforce well-being and a commitment to the health of the nation:

- as an expression of corporate commitment to the care of staff;
- as part of a benefits package;
- a desire to be seen as a 'good employer'.

These reasons fall under the umbrella of 'health promotion is a good thing' and may engender a feeling of paternalism in senior management. This seems to be an additional reason for the paucity of follow-up studies, since it is the process, not the outcome, which is seen as important.

Workplace initiatives

The possible reasons why an organisation should consider health promotion initiatives have been fully discussed above. It is easy to target a workforce as it is a captive population. It is also convenient for the individual employee. The workforce constitutes a good age profile for health promotion and it is of economic importance to protect this group.

However, consideration must be given to appropriate health promotion topics in any one workplace. The health promotion needs of the workforce may be ill perceived and inappropriate (see Cholesterol screening, page 132). Certain criteria should be used in deciding on workplace initiatives:

- Action in the workplace must be appropriate. Certain areas may be seen by employees as constituting unwarranted interference in their personal lives.
- The initiatives should complement or be more efficient than those in the community. Government initiatives in recent years have much enhanced health promotion and screening activities in general practice. There is no point in duplicating these activities.
- The initiatives must complement activities concerned with the promotion of health at work. For example, a reduction in alcohol consumption may form part of an accident prevention programme.
- There are opportunities for collaboration with outside agencies such as the Health Education Authority (HEA). This maximises the use of available health promotion material. The HEA's Look After Your Heart campaign relies heavily on joint working relationships.
- The message to be conveyed should be clear and uncontroversial. The continuing debate about the significance of moderately raised cholesterol levels provides a good example of a confused message.
- It must be possible to set targets and measure outcomes.

Scope for workplace initiatives

There are three types of intervention: screening, education and creating a healthy environment. Topics commonly considered as appropriate for workplace health promotion are:

- coronary heart disease;
- smoking;
- alcohol;
- stress;
- cancer.

Criteria for screening in the workplace should be narrow. The factor chosen should:

- be appropriate to the working age population;
- be appropriate to the particular workforce;
- have clear boundaries of normality;
- have effective treatment available;
- have screening techniques which are acceptable in the workplace and not disruptive;
- not be easily available elsewhere.

Education and screening

Coronary heart disease

In the UK, coronary heart disease accounts for 26 per cent of deaths. This is one of the highest levels in the world, the single largest cause of death and the single main cause of premature death.

Strokes, which have similar risk factors, account for a further 12 per cent of deaths. The government objective is to reduce deaths from coronary heart disease by 30 per cent by the year 2000 and to reduce deaths from strokes by 40 per cent.

Risk factors known to be associated with these disorders include:

- obesity;
- hypertension;
- high cholesterol;
- stress;
- smoking;
- excessive alcohol consumption;
- genetic factors.

All these, except the last, are amenable to health education.

A major contribution to health promotion in relation to coronary heart disease is the HEA's Look After Your Heart (LAYH) campaign, launched in 1987. Its workplace initiative requires employers to sign up for three out of a list of ten initiatives:

- provide information on health issues;
- develop a comprehensive smoking policy;
- provide and promote healthy food choices;
- introduce physical activity programmes;

- promote sensible drinking;
- identify sources of stress and provide staff support;
- provide health checks;
- provide an environment conducive to health;
- use LAYH workplace services;
- explore new initiatives on health.

To date, the programme is available to 3.5 million employees. Becoming part of this campaign is usually a high-profile activity and the HEA expects to have support from the highest level of management in the organisation. The HEA provides promotional material and expert advice, and this may be particularly useful for smaller companies which do not have in-house expertise. In larger companies the initial commitment may be followed by related activities within the organisation, including fitness assessment, exercise classes, blood pressure screening, improved dietary choices in staff restaurants, slimming clubs and running clubs.

Fitness
There are various fitness assessment packages available. These generally measure heart and lung capacity by measuring pulse rates before and after some form of controlled activity, such as a step test or the use of the bicycle ergometer. Most of these packages include a basic lifestyle questionnaire and provide a useful printout on the individual's current state of health, with comments and advice on lifestyle and possible improvements.

Organisations offering fitness assessment to their employees often find it beneficial to provide exercise facilities, such as group membership of a local health club; an in-house gymnasium (occasionally a swimming pool); exercise classes; individual exercise packages; formation of running clubs, etc.

Blood pressure screening
It is common for individuals to be unaware that they have raised blood pressure (hypertension) because until high levels are reached the condition may be symptom-free. Hypertension is easy to treat and control; uncontrolled, it may lead to coronary heart disease and strokes. Identifying raised blood pressure is, therefore, an important health promotion activity. It is simple to carry out, takes minimum time, causes little disruption and can be undertaken by nurses. Apart from the occasional individual who may become anxious because a raised blood pressure is detected, the activity seems to be wholly beneficial.

Cholesterol screening

The case for cholesterol screening has not yet been established. Individuals with high cholesterol levels (above 8 mmol/L) have been shown to benefit from dietary and other therapeutic interventions, resulting in a lower incidence of coronary heart disease than those untreated. Individuals with moderately raised cholesterol have not been shown to improve their prognosis by attempts at reduction. Nevertheless, there is considerable public pressure to determine cholesterol levels. There is no doubt that a number of organisations continue to offer cholesterol screening as a sort of loss leader to enhance other more worthwhile health promotion initiatives.

Improved diet

There are still many aspects of the average diet which are less than satisfactory. The Committee on Medical Aspects of Food Policy (COMA) report showed that, as a nation, we consume food containing too much salt, unrefined sugar and fat, and too little fibre. The average percentage of food energy derived by the population from saturated fatty acids is 17 per cent and from total fat 40 per cent. Eight per cent of men are obese and 12 per cent of women. Of course, the individual must have a choice and staff restaurant menus cannot reasonably exclude, for example, chips. However, healthy choices should always be available and labelled. Many organisations use the traffic light system for labelling menu items, with red indicating the high fat content food and green the other extreme.

The BBC in Northern Ireland showed a decrease in the consumption of white bread, butter, non-fibre cereals, cream and dressed salads over a period of about two years when healthy alternatives were introduced in its staff restaurants. Gradual change to a healthier lifestyle can have a major effect on long-term national health.

Smoking

There is no doubt that smoking damages your health. National initiatives are limited to activities such as National No-Smoking Day. Audio-visual material is available from a number of sources such as Action on Smoking and Health (ASH). However, those who attend film sessions are frequently non-smokers. There can be few employees who do not know that smoking damages health. The emphasis in health education is, therefore, towards personalising the message and helping smokers to control the habit. This is discussed in full in Chapter 7.

Alcohol

Excessive alcohol consumption and its effects remain a largely hidden problem in the workplace. As with smoking, most people know that too much alcohol is bad for them (although a glass of wine a day appears to be beneficial!). The unit measurement of consumption has also had wide acknowledgement. However, a recent survey showed that 28 per cent of men and 11 per cent of women admitted to drinking more than 21 units a week. The social nature of alcohol consumption allows individuals to ignore the potential long-term ill effects and accept the obvious short-term problems (hangovers). Getting drunk, or at least heavy drinking, is still acceptable in many social circles and many work groups.

In the workplace it seems to be becoming less acceptable; as with smoking, one hopes it will soon become a social anathema. An approach to controlling and reducing the effects of alcohol consumption in the workplace is offered in Chapter 4.

Stress

In recent years it has been increasingly acknowledged that people experience stress in the workplace, and that this is unacceptable. It is difficult to know whether there has been an increase in the stress experienced or whether employees' expectations of a healthy workplace have increased. Because stress has become a popular topic, the available interventions have increased, and it may be difficult for the personnel manager to assess the efficiency, or indeed the legitimacy, of the facilities on offer. It is therefore important to make a proper assessment of the need for intervention. This is discussed fully in Chapter 4.

Cancer

Cancer is responsible for 25 per cent of all deaths. There is no doubt that early detection much improves individual prognosis. Screening for early signs of cancer has been explored in only a few conditions, notably breast, cervical, intestinal, lung and testicular cancer.

The European Code Against Cancer – Ten Commandments can be usefully publicised in the workplace:

- Do not smoke. Smokers – stop as quickly as possible and do not smoke in the presence of others.
- Moderate your consumption of alcoholic drinks – beer, wines and spirits.
- Avoid excessive exposure to the sun.

- Follow health and safety instructions at work concerning production, handling or use of any substance which may cause cancer.
- Eat fresh fruit and vegetables frequently, and cereals with a high-fibre content.
- Avoid becoming overweight, and limit your intake of fatty foods.
- See a doctor if you notice a lump or a change in a mole, or abnormal bleeding.
- See a doctor if you have persistent problems, such as a cough, hoarseness, a change in bowel habits or an unexplained weight loss.
- Women: have a cervical smear regularly.
- Women: check your breasts regularly, and if possible undergo mammography at regular intervals above the age of 50.

Lung cancer

There are 26,000 deaths from lung cancer each year. At a time when pulmonary tuberculosis was still a major cause of mortality, mass radiography for the population was considered to be economically sensible and significant in terms of public health. However, with the decline of tuberculosis, the percentage of chest X-rays revealing unsuspected abnormalities fell to less than 0.1 per cent. The simultaneous growth of awareness of the possible harmful effects of repeated X-rays led to the end of mass radiography. Unfortunately, lung cancer can be far advanced before symptoms develop and the thrust of activity against this disease now largely lies in efforts towards smoking reduction. Ninety per cent of lung cancers occur in smokers. There may be something to be gained by increasing public recognition of unacceptable symptoms, such as a persistent cough and coughing up blood, but it is surprising how long symptoms can be ignored.

Colonic cancer

Research continues into a practical early test for colonic cancer. It is well established that traces (not visible) of blood appear in the stools at an early stage of this disease. Various tests have been developed to detect this. But to date these have proved difficult to use and also not entirely specific, providing false negatives and false positives. Colonic cancer is responsible for over 6500 deaths per annum and the search for a suitable screening test continues.

Testicular cancer

Testicular cancer has been particularly related to men who had local contact with mineral oils, such as light engineering workers. These cancers have now been almost totally controlled by improved work practice. However, testicular cancer of unknown origin occurs in a significant number of young men each year. As with the breasts, self-examination of the testicles is simple and effective, and should be encouraged. Information and advice is best distributed by means of pamphlets and advice sheets, which can be obtained from organisations such as the Cancer Education Co-ordinating Group.

Cervical cancer

Cervical cancer screening continues to be high on the public agenda. Cervical cancer is responsible for over 2000 deaths per annum. This number has decreased consistently since the introduction of general screening programmes in the 1980s. At present women between 20 and 64 years of age are offered screening every three years. Unfortunately, the extent of the screening programme has led to some spectacular disasters, either in inaccurate reporting on the cytology, failure of communication of the results, or unsupervised and unsatisfactory screening procedures. Nevertheless, the overall programme has proved successful in early detection, and therefore better treatment results, in a significant number of women.

Inevitably, the downside of a programme like this is that, in a small number of cases, the individual may have fears raised unnecessarily, but this is a small price to pay for a reduction in deaths from the disease. In the 1980s many organisations (notably Marks & Spencer) provided cervical cancer screening in the workplace. This supplemented the vestigial screening programme available in the general practice setting and seemed to reach women who otherwise might not have bothered to have the test. However, the programme for cervical cancer screening now available in the community has largely supplanted workplace initiatives. The cost to employers may be considerable (about £8–10 simply for reading the smear). Added to this is the cost of staff time to perform the test and of the individual being tested. It is unlikely that there is any substantial financial gain to employers.

It is, nevertheless, useful to create awareness of the benefits of screening and provide information on the process as part of general health education. Information leaflets can be obtained from organisations such as Tenovus (see Useful Addresses, pages 211–15).

Breast cancer

Breast cancer screening using radiological techniques (mammography) is available to all women between 50 and 64 on a three-year basis. It still remains the major cause of death in women in their middle years and is responsible for around 15,000 deaths per annum. It is usual for breast examination to be carried out by the operator at the time of cervical cancer screening. Self-examination of the breasts has been well developed and is practised by many women. There has, however, been some controversy about its efficiency.

It is probably inappropriate for an organisation to provide screening for breast cancer in the workplace as facilities are now generally available in the community. But, as part of a health promotion campaign, it is useful to provide leaflets and audio-visual material to enhance awareness and demonstrate self-examination techniques.

Glaucoma
Like hypertension, glaucoma (increased pressure within the eye) may reach an advanced stage without producing any symptoms.

Without treatment it can ultimately result in blindness; with treatment it can be controlled and any loss of visual acuity avoided. There is, therefore, considerable benefit in screening programmes which can detect early increases in ocular pressure. Inevitably, these techniques require special skills and also expensive apparatus. Nevertheless, there have been a number of successful screening programmes in the workplace.

Because of the skills required, it seems unlikely that many employers would be prepared to support this type of screening programme.

Action plan

Having once made the commitment to health promotion in the workplace, it is necessary to draw up some sort of plan to put that commitment into action. The plan should consist of five parts:

- assessing the needs of the organisation;
- defining appropriate target areas;
- deciding who will deliver the health promotion;
- developing a programme;
- assessing the resource implications.

Assessing health promotion needs

The demographic breakdown of staff will be particularly important in deciding appropriate target areas. Clearly, a workforce which is predominantly male and engaged in manual tasks will require different interventions from a female clerical workforce. There are also geographical variations in behaviour, and what may be appropriate in a northern mining community (if such still exists) may well not be so in a southern seaside town.

The sophistication of the workforce with regard to knowledge of health facts is also an important factor in determining the type of intervention that will be appropriate. In areas such as smoking and alcohol consumption it may be less necessary to give facts in relation to their own behaviour. The assessment process may require a survey of attitudes, knowledge and expectations. A pilot study to measure parameters, such as levels of stress, obesity and current use of screening facilities, may also help to clarify appropriate target areas. The assessment should also determine what facilities and health promotion activities are available in the community.

Defining target areas

At the end of the assessment period it should be clear which areas of health promotion are most suitable for your workforce within the community setting. The possible areas of health promotion have already been fully described. You may decide on several different interventions addressed to different parts of the workforce, or one overall target area. You should have decided what messages you want to get across. For example, if smoking is still not controlled within the working day, a health promotion campaign to highlight the adverse effects of smoking may precede the development and implementation of a non-smoking policy.

Obtaining advice

At this point some consideration should be given to whether there is sufficient expertise within the organisation or whether outside experts need to be approached. The Health Education Authority will provide information and limited help. Local authority initiatives should be explored through the health promotion officer of the appropriate authority. If these experts cannot help, they will be able to suggest other voluntary bodies (see Useful Addresses, pages 211–15).

Developing a programme

The health message can be delivered in a number of ways:

- one-to-one counselling;
- screening;

- exhibitions;
- posters;
- national days;
- workshops;
- seminars;
- audio-visual material;
- catering initiatives;
- health-related personnel policies.

There is no doubt that presenting the message in different ways over a period of time and, where appropriate, introducing some form of individual measurement are the most powerful methods of changing behaviour. An example of this approach is given at the end of this chapter (see page 140).

Learning is, of course, always enhanced by involvement of the target individual. It is not enough to provide information by means of leaflets and audio-visual material alone. Providing opportunities to monitor relevant individual concerns such as blood pressure, fitness, respiratory function and weight, which will inevitably be associated with discussion, can be effective.

Targeting the individual may be seen as an almost cynical exercise if other messages in the workplace are negative. Therefore, the creation of a healthy working environment should be part of any health promotion plan. This may include:

- a non-smoking policy,
- the restriction of alcohol availability;
- the provision of healthy food choices;
- the provision of exercise facilities;
- a safe workplace;
- healthy hours of work;
- opportunities to develop leisure interests.

Resource implications

These will vary enormously depending on the type of programme developed and the cost of any expert help. For any programme, apart from the most basic involving merely leaflets and posters, the cost of each employee's time is an inevitable component. If your programme forms part of a national or local initiative, there may be minimal additional costs. If it is not part of such a programme, costs will involve the production of promotional material and the use of experts.

There may be considerable costs incurred in the introduction of a smoking policy if structural alterations are required to provide limited accommodation for smokers.

A sample health promotion initiative

Nielsen Marketing Research

Nielsen signed the Look After Your Heart charter in 1987. They introduced the following:

Annual health checks are offered to staff. They include discussion of lifestyle; stress and physical fitness levels; blood pressure check; urine analysis; cholesterol testing; and check for anaemia.

Exhibitions and displays on healthy living are set up regularly in reception areas and at staff entrances.

There is a well-equipped sports and social centre, and opportunities are provided for staff to take exercise. There are keep-fit sessions, yoga classes, and facilities for tennis, hockey, football, rounders and cricket.

There are voluntary no-smoking agreements within departments (with no smoking between 10 am and 4 pm).

Articles on health appear regularly in the in-house newspaper.

Monthly stress management courses are available to all employees.

Future activities

The expansion of the LAYH workplace project to the other two company sites in the UK is planned.

Part III

Chapter 9
The employment of people with disabilities

> If society was organised on a more equitable basis, many of the
> problems associated with not being physically 'perfect' – as if such
> a concept had any logical basis – would disappear.
>
> (Simon Brisenden, *Disability, Handicap and Society*,
> 1986, Vol. 1, No. 2, p. 176)

There are nearly 2½ million people of working age in Britain who have disabilities. Fewer than one-third are in employment. British law requires employers with 20 or more employees to employ at least a 3 per cent quota of registered disabled people. Until recently, government policy has been to educate, persuade and increase awareness. Legislation to be enacted in Autumn 1995 makes unjustified discrimination on the grounds of disability unlawful. In this chapter the business case for employing people with disabilities is presented. The tendency for employers to equate disability with illness, and to present health and safety legislation as a barrier to the employment of more people with disabilities, is shown to be unjustified. Details of the help and support available to employers from such sources as the Employment Service are given. An outline policy statement is included for guidance.

Introduction

There are more than 6 million disabled adults in Britain: over 14 per cent of the adult population. Nearly 2½ million of those with disabilities are of working age but only 31 per cent are in employment.

Increasingly, it is being recognised that people with disabilities can make a full contribution to working life, yet they frequently suffer unfair discrimination in employment. This is not only morally wrong; it is bad for business and may be unlawful.

Among people with disabilities there is an enormous range of abilities, interests and personal attitudes. Ignorance, prejudice and management

practices often prevent the employment of such people, and so the
advantages which could accrue to the employer are lost.

The business case for employing people with disabilities

- 'People are the key to success in any organisation.'
- 'We wish to place on record the appreciation this company has for
  the people within the company, without whose contribution we
  would not have achieved this year's gratifying results.'
- 'People are our most important asset.'

These three statements – the first from an article in a leading
management journal; the second from the annual report of a leading
company; and the third from the Chairman of the Board when addressing
the company's Annual General Meeting – represent the importance that
commercial and industrial undertakings place on the human resources
within the organisation.

In today's highly competitive business environment, it must be a key
strategic objective to employ effective people capable of satisfying and,
in total quality terms, exceeding customer needs and expectations.

If people with disabilities are excluded from an objective assessment
of what they can contribute to the present and future needs of an enterprise in recruitment, retention, development or promotion, the company
is unnecessarily restricting its own ability to optimise each position in
the organisation.

All jobs require certain abilities and qualities, and it is the assessment
of a person's ability to meet the carefully analysed job specification which
is important.

People with disabilities do not necessarily have a job-related impairment,
and only a minority of cases need some adaptation of the environment. In
those cases where some adaptation is necessary, help with the costs may be
available (see pages 150–1). It can be argued that if the most sought-after
qualities in employees today are commitment, flexibility, effective
communication, good concentration, dealing positively with change and
the ability to recognise and use opportunities imaginatively, people with
disabilities, who will have had much practice in converting day-to-day
obstacles into opportunities, will have developed these skills to a greater
degree than their able-bodied colleagues.

Matching ability to job

There is a danger that people will be pigeon-holed rather than viewed
as unique individuals. To some, 'a person with disabilities' conjures up

a mental picture of someone who uses a wheelchair or who is blind. While there are such obvious disabilities, of course, many more are hidden, such as diabetes, dyslexia, a heart condition, depression, asthma and back problems.

It is recognised that there are occasions when people need to be viewed as a group or a type; for example, one such group might be those requiring a sedentary occupation, while another might be those who should not come into contact with irritant materials because of a skin condition.

In the field of employment, however, consideration should always be given to an individual's ability. A person's ability to do his job – or his potential to do another collection of tasks which we call a job – is what is important.

Are there any jobs for which a person with diabetes, whose condition is controlled by insulin, should not be considered?

> Disability should be seen for what it is; the result of an interaction between each individual's physical or mental impairment and the environment in which he or she functions.
>
> In a properly adapted work situation the employee in a wheelchair will not be disadvantaged at all. It is the barriers imposed by the environment which become the disabling factor. These barriers are not just physical, they are also attitudinal. Recognising this can go a long way towards removing the stereotypical image which can so easily get in the way when you are assessing a person's ability to do a job.
>
> (Employers' Forum on Disability 1992:5)

The concept of integration is one which is widely supported in the field of employing people with disabilities. The idea of a special or separate group within the workforce is (with the obvious exception of those employed under the sheltered employment or sheltered placement schemes) not in keeping with contemporary thinking. This emphasises the need to take into account, and cater for, an objective assessment and matching of job needs with the presence or absence of related and necessary abilities.

Disability awareness

As has been stressed, objective assessment of job needs and the match or mismatch of an individual's abilities are fundamental. As most recruitment, retention, promotion and development decisions are, in all but the smaller companies, taken with the involvement of representatives from different functions – line management and personnel staff being typical – it is strongly recommended that those involved in these decisions receive disability awareness training. Such training is not only designed to assist with objective assessments and judgements, but also to help those interviewing in disability 'etiquette'; it is only natural that those who are inexperienced in this field are concerned about what they say or do.

Disability etiquette is the concept of a code of behaviour which supports politeness and the maintenance of dignity when dealing with people with disabilities. One example is to try to avoid using the phrase 'restricted to a wheelchair', when it is a wheelchair that gives the user his or her degree of freedom. Disability etiquette suggests that 'user of a wheelchair' is the preferred phrase. A course designed to meet training needs in disability awareness would typically focus on such aspects as what it feels like to have a disability, special aids to employment, and how a number of specific conditions have an impact on employment. Courses are available which concentrate on a specific condition. One example is deafness, where the different types of deafness and how these affect the acquisition of speech and language are explained. Attendees may be given some instruction in basic communication skills, sign language, finger spelling and lip reading, as well as an introduction to the technical aids available.

In formulating personnel procedures, consideration has to be given to arrangements for the sensitive collection of information, application form design and reception and interviewing facilities being among the most obvious.

Legal considerations

At the present time there are two parliamentary bills at committee stage – The Governments Disability Discrimination Bill and the Civil Rights (Disabled Persons) Bill – 'The Barnes Bill'. The differences between the two bills include the nature of the agency to police the new laws, the Governments Bill proposing the establishment of a National Disability Council to advise the minister for disabled people, the Barnes Bill proposing a Disability Rights Commission, independent of government and responsible to parliament. The definition of disability and the size, in employee number terms, of the firms to which the measures would be applied also differ.

The Disabled Persons (Employment) Acts 1944 and 1958

The Disabled Persons (Employment) Act 1944, as amended by the 1958 Act, established a voluntary register of people with disabilities. The 1944 Act placed certain duties and obligations on employers with 20 or more workers (including those working at any branch), relating to the employment of people with disabilities who are registered under the Acts.

Quota scheme

Under the quota scheme, employers with 20 or more workers have a duty to employ a quota of registered disabled people. The standard quota is 3 per cent of an employer's total workforce. It is not an offence to be below quota, but in this situation an employer has a further duty to engage suitable registered disabled people, if any are available when vacancies arise.

An employer who is below quota must not engage anyone other than a registered disabled person without first obtaining a permit to do so from the local Jobcentre. An employer must also not discharge a registered disabled person without reasonable cause if he or she is below quota as a result.

A permit may be issued either to enable employers to fill immediate current vacancies if there are no suitable registered disabled people available, or, as a bulk permit, to authorise in advance the engagement of a specified number of workers during a period of up to six months. Bulk permits are issued on the understanding that employers will notify details of vacancies which arise to the local Jobcentre, and that they will consider positively the engagement of any suitable registered disabled people who may become available.

Designated employment schemes

The 1944 Act also included the power to reserve certain occupations for people with disabilities. Entry into the occupations of car park attendant and passenger electric lift attendant is reserved for registered disabled people. Registered disabled people employed in such occupations do not count towards an employer's quota.

Record-keeping

Employers who are subject to quota or who have designated employment are required to keep records showing the total number and names of all people employed, together with the dates of their starting and finishing employment. These records must identify, among others, the following employees:

- registered disabled people, including those whose registration has lapsed during their employment with the employer;
- a person employed under a permit;
- a person employed in designated employment.

These records must be produced for inspection by officials authorised for the purpose by the Employment Service.

The Companies Act 1985

Section 235 and Part III of Schedule 7 of the Companies Act 1985 are administered by the Department of Trade and Industry.

The statutory requirements apply to the directors' report of all companies, as defined in the Companies Act, employing on average more than 250 people. For these purposes no account is taken of persons working wholly or mainly outside the UK.

The directors' report of these companies must contain a statement describing the policy applied during the previous financial year:

- for giving full and fair consideration to people with disabilities applying for jobs, having regard to their particular aptitudes and abilities;
- for continuing the employment of employees who become disabled while working for the company and for arranging training for them if appropriate; and
- for the training, career development and promotion of employees with disabilities.

Although the statutory requirements do not apply to public sector employers, the obligations which the requirements impose fall equally on them. These statutory requirements relate to policies towards the employment of workers with disabilities generally and not only, as in the quota scheme, to those who are registered disabled under the Disabled Persons (Employment) Acts.

The Health and Safety at Work etc. Act 1974

This Act imposes a general duty of care towards all employees. The implication is that an employer must pay particular attention to the needs of people with disabilities and, if appropriate, monitor at regular intervals their suitability for the work on which they are employed.

The Management of Health and Safety at Work Regulations 1992

These are the second of six sets of regulations which stem from recent European Community Directives, and came into force on 1 January 1993. In the context of this chapter the most significant regulation is in relation to risk assessment. Employers are required to undertake assessment of all risks to health in individual workplaces, and in addition the approved Code of Practice requires the employer to identify groups of workers who may be particularly at risk, and as one of the examples of such a group cites 'any disabled staff'. The regulations also require employers to record the significant findings of their risk assessment, which should include the population which may be at risk and any groups of employees who are especially at risk. The Code stresses that the purpose of risk assessment is to help the employer or self-employed person to determine what measures should be taken to comply with the employer's or self-employed person's duties under the relevant statutory provisions.

The Workplace (Health, Safety and Welfare) Regulations 1992

These regulations, together with an approved Code of Practice, come into effect in two stages. Workplaces used for the first time after 31 December 1992 have to comply as soon as they are in use. In existing workplaces, apart from any modifications, the regulations take effect on 1 January 1996.

The approved Code of Practice contains additional guidance which specifically requires that workplaces must meet the health, safety and welfare needs of each member of the workforce, including people with disabilities. Such aspects as traffic routes, facilities and workstations are singled out for attention.

Access, means of escape and refuges are all mentioned, and references are made to the relevant Building Regulations and British Standards. There is a recognition that, in the case of an existing building, compliance with the Code is not always possible, but it is recommended that alternative ways of meeting the objectives of the Code should be sought and it is stressed that non-compliance with recommendations should not be regarded as adequate grounds for excluding people with disabilities from a building.

Employment service initiatives

As previously stated, the case for integration has, by and large, been accepted. The Employment Service, therefore, encourages optimum use of its mainstream provisions, i.e. Jobcentres, Training for Work, Youth Training and Job Clubs etc.

In 1992 the special disability services were reorganised. There are now 70 fully operational local specialist teams. They are required to deliver services for employers and people with disabilities and are known as PACTs (Placing Assessment and Counselling Teams). The team members are called Disability Employment Advisers (DEAs). PACTs can give:

- advice and help on recruitment and training needs, and on retaining people who become disabled;
- information on special schemes and services to help in the employment of people with disabilities;
- advice to managers on working successfully with employees with disabilities;
- advice on obligations under the quota scheme and permits when appropriate;
- advice to people with disabilities who need specialist help to obtain employment;
- assessment and counselling designed to identify a person's abilities and aptitudes, and how these can best be used or developed;
- information to employers about their clients, both speculatively and in response to notified vacancies.

PACTs also contract with a wide range of organisations which provide work preparation and training for people with disabilities.

Strategically placed around the country are Ability Development Centres (ADCs) whose role is to offer enhanced assessment and specialist advice on most aspects including disability awareness.

There is also the Major Organisations Development Unit (MODU) which offers a special good practice advisory service on the employment of people with disabilities to large employers.

From June 1994 the Access to Work Scheme has provided financial assistance to the employer and/or the person with the disability, up to a value of £21,000 over five years. Access to Work can pay for:

- a communicator for people who are deaf or who have a hearing impairment;
- a part-time reader or assistance at work for someone who is blind;

- a support worker if someone needs practical help, either at work or getting to work;
- equipment (or adaptations to existing equipment) to suit individual needs;
- adaptations to a car, or taxi fares or other transport costs if someone cannot use public transport to get to work;
- alterations to premises or a working environment so that an employee with a disability can work there.

People who have a long-term illness or disability and who are at a disadvantage in getting a job may receive a Disability Working Allowance. This is not strictly an Employment Service initiative as it is administered by the Department of Social Security. The allowance is designed to help people to return to, or take up, work even though they may be unable to earn enough to replace the benefit they are entitled to while they are incapable of work. It is generally available, therefore, to people in receipt of social security benefits who take up low paid employment.

The disability symbol

The Employment Service has designed a symbol (see Figure 9.1) which is intended to be used by employers to show their commitment to good practice in employing people with disabilities, and to advise people with disabilities that the symbol-using employer will be positive about their abilities. It is to be used on company stationery, letterheads, application forms and job advertisements. Since June 1993 all employers using the symbol are required to make the following commitments to action:

- To interview all applicants with a disability who meet the minimum criteria for a job vacancy and consider them on their abilities.
- To ask disabled employees at least once a year what can be done to make sure that they develop their skills and use them at work.

Figure 9.1 The disability symbol

- To make every effort, when employees become disabled, to ensure that they stay in employment.
- To take action to ensure that key employees develop the awareness of disability needed to make the commitments work.
- To review these commitments and what has been achieved each year, plan ways to improve on them, and let all employees know about progress and future plans.

Current legislative developments

The government has introduced a bill and published proposals to bring in new laws and measures aimed at ending unjustifiable discrimination against people with disabilities.

It is their intention that it becomes law by Autumn 1995 when the 1994 act would be repealed and thus the quota and designated employment schemes would cease.

The new legislation on employment defines a person with disabilities as a person with a physical or mental impairment which has a substantial and long-term adverse effect on his or her ability to carry out normal day-to-day activities.

A long-term impairment will be defined as one that has lasted, or can reasonably be expected to last, for at least twelve months. When an impairment stops having a substantial adverse effect on a person's ability to carry out normal day-to-day activities, it will be treated as continuing to have that effect if the impairment is likely to recur. A mental impairment will include a mental illness if it is a clinically recognised condition. An impairment will be taken to mean normal day-to-day activities if it affects mobility, manual dexterity, physical co-ordination, continence, ability to lift, carry or move everyday objects, speech, hearing or eyesight, memory or ability to learn or understand, or perception of the risk of physical danger. The definition will also apply to those people whose disability is regulated by medication or by the use of a special aid.

People with severe disfigurements will not be required to satisfy the 'substantial effects' provision but the condition will be required to be 'long term'.

People with progressive conditions (for example, cancer, multiple sclerosis or muscular dystrophy) will be covered by the definition where the condition would be expected, in the future, to have a substantial effect on the person's ability to carry out normal day-to-day activities.

The right of non-discrimination will refer to all aspects of employment including recruitment, promotion, dismissal, training, etc. and will make it unlawful for an employer to treat a disabled person less favourably because of their disability without justifiable reason. The new right will not affect an employer's freedom to recruit and promote the best person for the job, nor will it affect employers' general duty to comply with laws on health and safety. Employers will be required to make a reasonable adjustment to working conditions or the workplace where that would help overcome the practical effects of a disability. In deciding whether an adjustment is reasonable, employers will be expected to take account of the costs of the adjustment and how effective it will be in overcoming the practical effects of a disability.

The right will apply to employees, job applicants, apprentices and people who contract personally to provide services and will cover both the public and private sectors.

A code of practice will be drawn up giving guidance on the new requirements and practical advice to help employers comply.

Disabled people who feel that they have been discriminated against unlawfully in employment will be able to complain to an industrial tribunal and seek the help of conciliation officers in resolving a dispute without a hearing.

Conclusion

Full and fair consideration for employment, retention, training and development should be the guiding principle. While illness may have caused some disabilities, a person with a disability is not necessarily ill.

Employers should produce a policy statement and establish objectives against which their practices and action plans can be monitored. This will help to avoid a situation where good intent remains just that. The following policy statement used at London Electricity is a good example of such a document.

A policy on the employment of people with disabilities

Objectives

The company accepts a social and moral obligation to play a significant role in providing for the employment of people with disabilities. We wish to be recognised by the community as an organisation which provides good employment opportunities to disabled people. People with disabilities who apply to us for jobs should know that they will receive fair treatment and be considered solely on their ability to do the job. We will strive to meet and exceed our legal obligations in employing people with disabilities.

Consultation

Our policy and practices on the employment of people with disabilities have been discussed at the Equal Opportunities Working Party consisting of management representatives and staff representatives from all negotiating bodies and trade union officials.

Responsibilities

The director of personnel is the executive responsible for the implementation and monitoring of the policy. His division will provide advice as required and will liaise with relevant outside organisations. Welfare and occupational health staff will provide specialist advice and help where necessary.

Managers and staff are responsible for carrying out the policy; for example, by implementation of the good practices outlined below.

Good practices

The company is committed to the following good practices.

Recruitment

Full and fair consideration will be given to disabled applicants for all types of vacancy. Supervisors should have an open mind about which jobs people with disabilities can do, even those with severe disabilities.

Training will be provided for managers and supervisors about the policy.

Employees who become disabled

Employees who become disabled will be dealt with in accordance with the policy on long-term sickness if this is appropriate.

Consideration will be given to special assistance to enable the employee to stay in the same job or take a new one; for example, special aids, adaptation to premises, modification of job content, part-time working, etc. Advice can be sought from the occupational health adviser and from the personnel division.

Special schemes

Government-sponsored schemes and facilities, such as the job introduction scheme, sheltered placement scheme and employment rehabilitation centres, will be used where this will assist in recruiting or retaining disabled staff.

Outside organisations

We will develop and strengthen links with outside organisations such as disablement employment advisers (DEAs), the Placement Assessment and Counselling Teams (PACTs) and voluntary organisations concerned with the employment of disabled people. Targets will be set for the recruitment of disabled people and progress against these targets will be monitored.

Monitoring

The company's policy and practices on the employment of disabled people will be reviewed regularly by the personnel division. Implementation of the policy will be monitored as part of the equal opportunities audit.

We will concentrate on disabled people's abilities rather than their disabilities when considering them for jobs.

All job applicants called for interview will be asked whether they require any special facilities or assistance at the interview.

We will be prepared to make minor modifications to the content of jobs to accommodate a disabled applicant.

Adaptations to premises will be made where necessary and practical.

We will use special aids or equipment where appropriate.

Disabled candidates who meet the essential criteria for a job will be short-listed for interview.

Career development

Staff with disabilities will be given equal opportunities for training, career development and promotion. We will seek to develop the skills and potential of disabled employees to the full.

Health and safety

Staff with disabilities and their colleagues will be fully briefed on what to do in the case of fire or emergency.

The normal procedure for completion of medical questionnaires and medical examinations will apply to disabled applicants. It is recognised that disability does not necessarily equate to ill health.

Training and publicity

Information about our policy and practices on the employment of people with disabilities will be included in the annual report.

Chapter 10
The employment of women

Unless career women organise help to handle the base camp, they are left to do everything – their jobs, be a parent, support husband and arrange family gatherings like Christmas.

(Siobhan Hamilton-Phillips, DPS Consultants Ltd, Newsletter, December 1993)

Women account for over 44 per cent of the people in paid employment. The numbers of women with dependent children and those remaining at work during, and for a greater part of, their pregnancy are increasing. Legislation has been enacted making discrimination on the grounds of sex unlawful and affording protection in such matters as pay and the promotion of equal opportunities. This legislation is discussed as are the implications for employers. The chapter highlights the dilemma thrown up when the need to provide a safe place for work conflicts with the individual's right to privacy in such matters as actual or intended pregnancy. Family-friendly policies are outlined. The change in the state retirement age for women is noted against the background of precedent-making judgements in respect of pension arrangements.

Introduction

There are now over 10½ million women aged between 16 and 59 in paid employment, which represents an increase of 1½ million in the last decade. Women account for over 44 per cent of those in paid employment. There has been a steady increase in the number of working women who have dependent children, and the number of single mothers has also steadily increased. Another trend is for women to remain at work during, and for a greater part of, their pregnancy.

As women have sought (and legislation has been enacted in the field of) equal opportunities, they have entered into traditionally male areas of work, although some 60 per cent of employed women are engaged in the clerical, educational, health and personal service sectors of the economy.

Legislation

In addition to general health, safety and welfare legislation there are other statutes which must be observed. The Sex Discrimination Act 1975 (as amended in 1986) and the Equal Pay Act 1970 are designed to remove discrimination and promote equal opportunities, while the Trade Union Reform and Employment Rights Act 1993 affords protection for women employees when they are pregnant.

The Equal Pay Act

The Equal Pay Act 1970 is concerned with the terms and conditions in a woman's contract of employment, which must clearly show the rate of remuneration she is, or is going to be, paid, and contains provisions designed to ensure equality in terms of pay. The Act defines three conditions which must be met for a woman to succeed in a claim for equal pay under the arrangements provided by the Act. They are:

- The male comparator must be employed at the same time.
- The male comparator must be employed by the same or an associated employer at the same establishment or at a different establishment belonging to the same or an associated employer, where common terms and conditions of employment are observed either generally or for relevant classes.
- The female must be engaged on like work or work rated as equivalent or work of equal value to that of the male comparator.

The Sex Discrimination Act

The Sex Discrimination Act 1975 (as amended in 1986) is designed to prevent unlawful discrimination against women, and against men on the same grounds or against a married person of either sex on the grounds of that person's marital status. Discrimination is permitted in those special arrangements emanating from pregnancy and childbirth.

There are three types of unlawful act to consider:

- *Direct discrimination*, which is where a person of one sex is treated less favourably than a person of another and the sex of that person is the reason for the unfavourable treatment.
- *Indirect discrimination*, which is where a person applies a condition or requirement to another, such that the proportion of persons

of one sex who will not be able to meet that condition or requirement is much larger than the other sex. Obviously, if the employer
is able to show that the condition or requirement is justified,
irrespective of sex, no indirect discrimination will have taken
place.

- *Victimisation*, which arises because the person has brought proceedings or has given evidence or information in connection with
proceedings under the Act or the Equal Pay Act 1970. It is also
unlawful to victimise a person because the individual has done
something in relation to either Act to any person, including
the discriminator, or has made allegations of a contravention of
either Act unless the allegation was false and not made in good
faith.

In the second part of the Act five types of unlawful discriminatory act
are specified and are concerned with:

- The arrangements a person makes for the purpose of determining
who shall be employed. These arrangements have to ensure that
job opportunities are available to all, irrespective of sex.
- The terms on which a person offers employment to another. There
are exceptions, one of which is pay, covered by the Equal Pay Act
1970, and differences where the employer can show that these are
to do with genuine material difference between the two applicants
which has nothing to do with their sex.
- Refusing or deliberately omitting to offer employment because of
a person's sex.
- The way a person offers access to opportunities for promotion,
transfer or training, or to any other benefits, facilities or services, or
refuses or deliberately omits to afford access to those opportunities.
- Dismissing a person or subjecting him or her to any other detriment.
While it is obvious that if in a redundancy situation a person
is selected for dismissal by reason of sex an unlawful discriminatory act has occurred, the question of detriment is complex. If
a person is being subjected to unpleasant treatment of a sexual
nature, sexual harassment, this may or may not amount to a
detriment. The individual circumstances such as whether or not a
person of the other sex would have been vulnerable to the same
treatment and the degree to which there has been injury to the
person's feelings have to be taken into account in arriving at a
conclusion.

The Trade Union Reform and Employment Rights Act

The Trade Union Reform and Employment Rights Act 1993 affords individual employment rights which include those relating to maternity leave, maternity pay and the right to return to work as well as protection against unfair maternity dismissal. Every woman employee is entitled to 14 weeks' statutory maternity leave subject only to:

- giving the employer a written notice 21 days before her maternity leave period begins, stating that she is pregnant and giving the expected week of childbirth;
- giving the employer notice again at least 21 days in advance of the date on which she intends her maternity leave period to begin;
- producing medical evidence in the form of a certificate from a registered medical practitioner or midwife stating the expected week of childbirth if the employer requests this.

If an employee wishes to return to work before the end of her maternity leave period, she must give her employer at least seven days' notice of her intended return date.

If the employee intends to return to work she must include in her written notice of her pregnancy a statement that she intends to return to work, and she must give 21 days' notice of return at the end of her leave. During the statutory maternity leave period the employee will be entitled to all her contractual benefits apart from pay; a new statutory maternity pay scheme is introduced under this Act.

Dismissal or selection for redundancy on pregnancy- and maternity-related grounds is automatically unfair; there are no qualifying service conditions and women dismissed while pregnant or during their statutory maternity leave period will automatically be entitled to written reasons for dismissal. Women who would otherwise be suspended from work on health and safety grounds have to be offered suitable alternative work. If suspended, they will be entitled to their normal pay.

Currently, to qualify for the right to return to work up to 29 weeks after childbirth, the employee must have worked for the employer for at least two years at least 16 hours per week or for five years at least eight hours per week. A recent court ruling suggests that qualifications in respect of part-time work could be in breach of European Court rulings.

Under the terms of European law, Article 119 of the Treaty of Rome provides that: 'Each member state shall maintain the application of the principle that men and women should receive equal pay for equal work.'

The UK government appears reluctant to accept European Directives on other women's issues and has opted out of the Social Chapter under the Maastricht Treaty.

Other areas, such as pension rights and retirement ages, have been subject to decisions in the European Court, principally in the Barber case (*Barber v. Guardian Royal Exchange* 1990) which established that from 17 May 1990 sex equality in pension matters had to apply, while the Ten Oever case (*Ten Oever v. Stichting* 1993) reaffirmed the principle of sex equality and ruled on the meaning of the non-retrospective provision in the Barber judgment. The significance of this is that benefits need to be equalised only in respect of post-17 May 1990 service and that is the date from which spouses' benefits must be equalised.

The British government has recently announced that the state pension age will be equalised at 65, the equalisation to be introduced progressively between the years 2010 and 2020.

Health and safety

The European Commission has adopted a directive aimed at improving the health and safety of pregnant workers, those who have recently given birth and those who are breast-feeding. The Management of Health and Safety at Work Regulations have been amended accordingly (Regulation 13A-C). Under this comprehensive amendment a special risk assessment is required with respect to women who fulfil the above conditions. Where a risk is demonstrated, the hours of work or working conditions should be altered to avoid the risk where it is reasonable to do so. If it is not possible to remove the risk and alternative work is not available then the employee must be suspended with continued remuneration. Annex 1 to the Council Directive provides a list of agents, working conditions and processes that an employer must assess for risk; Annex 2 gives a list of agents and working conditions to which pregnant workers must not be exposed.

The hazards which must be assessed may be chemical, physical or microbiological.

Chemical hazards

The area which has probably caused most concern is that of chemical hazard to reproduction. It is perhaps important to say here that the vulnerability of the male reproductive system cannot be ignored. For example, infertility was demonstrated in male pesticide workers exposed to the chemical dibromochloropropane. It has been suggested that there is a higher incidence of leukaemia in the children of fathers working in nuclear processing plants.

It should also be noted that exposure to chemicals outside the work situation through alcohol consumption and smoking has been clearly associated with increased risk of miscarriage.

Several groups of chemicals have been implicated in causing reproductive impairment.

Lead

In 1989 the UK Factory Inspectorate reported that women exposed to lead at work were more likely to miscarry where there was significant environmental contamination and poor standards of hygiene. It is now known that children and adult females are more sensitive to lead exposure. In recognition of this, the Lead Regulations require that any woman of reproductive capacity should be withdrawn from work which exposes her to lead when the lead level in her blood reaches 40 ug/100ml or above. In contrast, men may continue to work up to a 70ug/100ml blood level.

Sterilising agents

Ethylene oxide is a fumigant for food stuffs and textiles and has been used in the sterilisation of surgical instruments. Accidental exposure has been shown to cause chromosomal damage. One study showed an increased incidence of abortions among hospital staff using ethylene oxide to sterilise instruments (Hemminki 1985).

Pregnant women should not be involved in work using ethylene oxide unless the process is totally enclosed.

Cytotoxic drugs

There is now widespread use of these drugs in the treatment of cancer. Safe working practices for administration and reconstitution only within laminar flow cupboards reduce risk to acceptable levels. There is no restriction on pregnant nurses being involved in administration under controlled circumstances. Most health trusts, however, restrict their involvement in reconstitution during the period of pregnancy and breastfeeding.

Anaesthetics

Several studies seemed to show an increase in spontaneous abortion among theatre staff exposed to low concentrations of anaesthetic gases. However, a review of 14 epidemiological studies failed to show any increased risk.

Physical hazards

Ionising radiation
This is used increasingly in industrial processes. The UK Ionising Radiation Regulations set specific limits of exposure for women of reproductive capacity and pregnant women.

Non-ionising radiation
This type of radiation includes radio waves and ultraviolet light. Attention has also focused on the small amount of electromagnetic radiation emanating from visual display units (VDUs). Several studies have shown no evidence that there is any increase in the number of miscarriages and birth defects among women working with VDUs. Studies by the National Radiological Protection Board confirm that levels of radiation are well below recommended acceptable levels.

Noise
There is a suggestion that exposure of the mother during pregnancy to high levels of noise may result in high frequency hearing loss in the child. However, studies have so far proved inconclusive.

Exercise
The increased number of women in the armed forces and their assumption of a full role has shown that high degrees of physical exercise may result in the transitory loss of menstruation. This has also been seen in ballet dancers. There appear to be no long-term effects. However, women are more at risk of musculoskeletal damage during pregnancy.

Microbiological hazards

Chicken pox, rubella and cytomegalovirus infection during pregnancy can have an adverse effect on the foetus. Pregnant women should avoid exposure to these infections. Clearly, those engaged in work with children or health care are more likely to be at risk.

Reducing the risk

As we have seen, in compliance with the Health and Safety at Work etc. Act 1974 the employer is required to provide a safe place of work as far as is reasonably practicable. This was taken to mean that it should be safe for the majority of workers. However, the new legislation in relation to pregnant women requires the workplace to be safe for them as a special group.

Do employers, therefore, have the right to exclude from the workplace those individuals who are particularly vulnerable for whatever reason? In situations which may be hazardous to women of reproductive age, as defined by the recent legislation, how is an employer to know of the pregnancy while still respecting the individual's privacy?

The Health and Safety Commission defines a woman of reproductive age as 'any woman capable of conceiving and carrying a child'. It suggests that any woman should be presumed to have reproductive capacity unless she provides a medical certificate stating otherwise. This could, of course, have wide-ranging implications for the employment of women. The Equal Opportunities Commission (EOC) has stated that acceptable levels of toxic exposure may need to be lower for women of reproductive capacity. To achieve this in a reasonably practicable way, organisations can either:

- suspend the employee on medical grounds;
- offer alternative work;
- dismiss the employee;
- give the employee the option of remaining, having explained the risks.

Any of these solutions may leave the organisation open to a claim of sexual discrimination. Alternatively, any subsequent damage to the foetus, if preventive action is not taken, may result in litigation and even prosecution under health and safety legislation.

The EOC has also indicated that questions about pregnancy and related matters on pre-employment questionnaires may be seen as discriminatory. Employers may well be caught between a rock and a hard place in their efforts to achieve fairness in the employment of women. At present, there is little case law to assist.

Family-friendly policies

Working women find it difficult to delegate domestic and family responsibilities. While social change is taking place, the norm is still for the woman to accept the greater portion of responsibility for the day-to-day management of the home, children, husband, and sometimes ageing relatives or grandchildren.

Company policies for women should recognise that, unless family-friendly policies are put into place, there is a danger that, in respect of women employees, performance may suffer and sick leave, absence and labour turnover increase.

In *Corporate Culture and Caring* (1993), produced for Business in the Community by the Institute of Personnel Management, family-friendly practices are said to enable an organisation to:

- make full use of the skills, experience and potential of all staff and thereby acquire the best person for the job;
- increase its efficiency, profitability and competitiveness;
- attract and retain committed, skilled and experienced staff, thereby recouping investment in training (and avoiding unnecessary recruitment-associated costs);
- increase the number of women who return to work following maternity leave;
- enhance the organisation's image with customers and staff through demonstrable commitment to equal opportunities.

The booklet gives examples of savings which can be made and an estimate of the costs which could be incurred in the provision of childcare support.

Initiatives may include options to provide childcare:

- workplace nurseries;
- partnerships with other organisations;
- private nursery places;
- sponsored child minders;
- childcare allowances and vouchers;
- after-school schemes;
- holiday play-schemes.

Another area for consideration is working arrangements and their flexibility. Flexible working can embrace:

- part-time working;
- job-sharing;
- flexible working hours;
- career or employment breaks;
- term-time working;
- flexi-place and teleworking.

All these should be supported by an effective information and advice service which specifically provides for dealing with crises.

Training and development

Business in the Community's Opportunity 2000 has been influential in gaining publicity for positive action designed to achieve genuine equal opportunities for women in employment. Training initiatives have been taken to facilitate the movement of women up the hierarchy into management positions and across into male-dominated occupations. While positive discrimination is unlawful in the UK, single-sex training designed to help women to realise their potential for management and male-dominated occupations is not. In terms of well-being, working women should understand the nature of stress and the human response to it, to recognise it in themselves and learn to manage the conflicting demands made on them.

Other types of women-oriented training and development programmes include those designed to address individual lack of confidence, especially in the areas of career potential and financial planning. There has also been an increase in help for women in order to improve their personal safety, particularly where they are employed in jobs exposing them to risk; for example, in occupations which provide a service to the general public and those involving driving, as in sales representative positions.

Conclusion

While this chapter has focused on the employment of women and described the issues and factors involved, employers are beginning to recognise that policies and practices initiated to benefit female employees, the lessons learned from them and the improvements made, can also profitably be extended to their male employees.

Part IV

Chapter 11
The use of display screen equipment

Write the vision, and make it plain upon the tables, that he may run that readeth it.

(Habakkuk)

The Health and Safety at Work etc. Act requires all employers to ensure the health and safety of their workforce. Special regulations apply to the use of display screen equipment, including any alphanumeric or graphic display screen. Employees covered by this legislation include those who use such equipment for an hour or more each day as a significant part of their work. This chapter outlines the requirements of the regulations. A checklist to facilitate assessment of the workstation is provided with guidance on its use. Simple arrangements for risk reduction are described in relation to basic ergonomics. Health issues such as repetitive strain injury and eye problems are considered in detail and guidance is given on avoiding health problems. At the end of the chapter a model chair-purchasing policy, a policy for eye and eyesight testing, and a policy for the implementation of the regulations are provided.

Introduction

The Health and Safety at Work etc. Act 1974 requires all employers to ensure the health and safety of their workforce. This is underpinned by regulations which are industry or hazard specific (for example, the Noise at Work Regulations). Recent legislation stemming from the EC Framework Directive applies across all employment sectors and is proving to be provocative, time-consuming and costly. A Directive places obligations on a member state to enact domestic legislation to achieve the requirements of the Directive within a stated time limit. In response to the Framework Directive, six sets of regulations (the so-called six pack) have been enacted under the Health and Safety at Work Act. The fifth of these is the Health and Safety (Display Screen Equipment) Regulations 1992. Guidance on implementation has been issued by the Health and Safety Commission. The aim of the regulations is to reduce the risk to staff of an occupational injury resulting from the use of display screen equipment.

Scope of the regulations

The scope of these regulations has been clearly defined and applies to any alphanumeric or graphic display screen regardless of the particular display process. The definition extends beyond the typical office visual display unit (VDU) and covers microfiche, lighted crystal displays and process equipment.

The workstation also falls within the regulations. The definition includes an assembly of display screen equipment (DSE) with or without keyboard or software, and any accessories to the equipment such as telephone, printer, document holder, work chair, work desk or work surface. It also includes the immediate workplace environment.

The term 'user' is applied to anyone for whom the use of DSE forms a significant part of his or her normal work. This has sometimes been difficult to interpret but it seems sensible to define a user as anyone to whom the following criteria apply:

- Their work often requires the use of a display screen for an hour or more.
- The display screen is used by them on most days.
- Their job could not be done without use of the display screen.
- The ability to use the display screen comes within their job specification.

Requirements of the regulations

Regulation 2 – Analysis of workstation and risk reduction

Employers are obliged to analyse each workstation (as defined above) in order to assess any risk to the user's health and safety. These assessments must be recorded. Having identified any risks, employers must reduce them to the lowest level reasonably practicable.

Regulation 3 – Timing of risk assessment and reduction

A workstation brought into use after 1 January 1993 must immediately meet the requirements. This includes workstations which are altered substantially or relocated. A workstation already in use on 1 January

1993 must meet the minimum requirements by 31 December 1996. However, they must be assessed as soon as possible after 1 January 1993 to determine any immediate significant risk to health.

Regulation 4 – Daily work and breaks

Employers must organise working practices so that work on DSE is regularly interrupted by breaks or other activities. It is becoming increasingly clear that, even with the ergonomically best possible workplace, health problems may arise if workers undertake long periods of work within a 24-hour period and do not take regular breaks.

Regulation 5 – Eye and eyesight testing

Employers must ensure that new employees are given the right to have an appropriate eye and eyesight test before commencing work. Employers must ensure that current users are given the right to have an appropriate eye and eyesight test on request.

Tests must be offered at regular intervals after starting work, and where a user experiences eye problems which it is reasonable to conclude are caused by working with DSE.

Employers must provide users with special corrective apparatus, i.e. spectacles suitable for the work being done, if their normal corrective appliances cannot be used and tests show that such provision is necessary.

Regulation 6 – Training for users

Employers must ensure that adequate health and safety training is given to an intended user before use of the workstation has commenced and to all users whenever the arrangements, including software, at their workstation are substantially modified.

Regulation 7 – Information for users

Employers must ensure that every user is given complete and comprehensible information about what has been done to meet the requirements of the regulations, about the entitlements to eye and eyesight testing, and training arrangements.

Assessing the workstation

A helpful first step in workstation assessment is the use of a questionnaire to be completed by users. The manager of each department should identify all workstations and users. Where there is significant variation between users and the time spent at workstations, it may be appropriate to start assessment with those who average over four hours daily at this work. A suitable questionnaire is given at the end of this chapter (see pages 180–3). The questionnaires should be returned to a designated person or department, depending on the size of the organisation and available skills. This may be the occupational health or safety or personnel department or, in the absence of these, the departmental manager. The completed questionnaires will indicate where there appear to be unsatisfactory features in the workplace, which should then be assessed by a competent person. In the absence of trained health and safety personnel, companies may decide to train staff to fulfil this role or may obtain advice from external consultants.

Factors to consider in risk assessment

The aim of risk assessment is to establish the situation with regard to the legal requirements for DSE (terminal and keyboard), the workstation, the environment and operational practices, and to discover current health problems.

Display screen equipment
Each terminal should:

- be adequately tested and guaranteed by the manufacturer as complying with relevant regulations;
- be maintained regularly;
- be provided with contrast and brightness controls;
- have an adjustable screen with the capacity to swivel and tilt;
- have easily read characters on screen without flicker or other distraction.

The keyboard should have non-reflective surfaces.

Workstation
The British Standards Institution has published standards for visual display terminals, visual display workstations and their design.

Equipment should be provided to enable the operator to adjust to the optimum working position. This should include a chair which is adjustable for height, has the facility to alter the height and tilt of the back support, and has a base of five feet on rollers. There are many suitable chairs available, but the variation in cost is enormous. It is therefore advisable to establish a policy on the central purchase of chairs (see the sample policy at the end of this chapter; pages 163–84). Users should be offered a small range of similarly priced chairs so that they can choose the most comfortable one for them. The hardness of the seat and the absence or presence of arms are the sort of variables which it is worth considering.

A document holder should be provided where appropriate. Bright and reflecting surfaces on adjacent furniture should be avoided. There should be adequate space (a minimum of 3.7 square metres per workstation). The work surface should be large enough to allow the comfortable arrangement of all equipment, and it should also be of a suitable height for DSE work with sufficient leg room. The use of a foot rest should be considered for short users.

Environment
Background noise should be kept to a minimum. Heating and ventilation should be kept at a suitable level.

Lights should be so arranged to avoid reflection and glare. General light levels should be 300–500 lux. In some cases where lower level general lighting is preferred, individual desk lights may be appropriate.

Workstations should be sited at right angles to windows and where necessary natural light should be controlled by blinds or curtains.

Operational practices
Long periods of intense, continuous work must be avoided by varying work tasks as well as by suitable rest and relaxation periods. This aspect of the legislation has probably been the most difficult to implement. Many jobs seem to require prolonged use of DSE. Natural breaks do not occur and if insisted on may cause more problems. Lack of rest breaks may be as much to do with the operator's persistence and conscientiousness as with the nature of the task. The intensity of work at the screen as well as its timing can be significant in the development of health problems. It is essential that managers control levels of screen work. A ten-minute break away from the screen every hour has been recommended.

Certain keyboard configurations such as QWERTY have been implicated in the development of work-related upper limb disorder (WRULD). However, many of the cases reported have been associated with particular repetitive work such as editing documents with continued use of the function keys. Software developments and proper training can help to avoid these repetitive activities.

There are so many tasks which need DSE that organisation of work may need to be considered on a job-by-job basis around what is reasonably practicable and what is operationally possible in order to avoid hazard to health. Strict interpretation of the suggested ten-minute break initially led to much confusion and industrial relations problems. However, common-sense seems to have prevailed.

Arrangements for risk reduction

The assessment, both individual and by a competent person, will have revealed where requirements are not being met.

Risk reduction must be implemented immediately where a significant risk has been identified, particularly if the user is already experiencing health problems. It should not be assumed that risk reduction measures will be costly. To a great extent, problems seem to arise because operators are not using their equipment correctly, do not have appropriate software and are not taking reasonable breaks. The single most common problem is inappropriate seating or seating used inappropriately. Figure 11.1 shows the correct position for working with DSE. Unsuitable positioning may develop because the height of the seat has not been, or cannot be, adjusted; the operator's feet do not touch the floor because he or she has short legs and there is no foot rest; the operator's vision is not suitably corrected for screen work; or there is not enough desk space. Risk reduction should be addressed in five ways:

- training in good practice;
- repositioning of equipment and chair adjustment;
- provision of equipment such as foot rest, document holder;
- appropriate lighting;
- work redesign.

Health issues

Discussion about the use of DSE and hazards to health is out of all proportion to the risk involved. It has been suggested that a number of conditions may arise from its use. These are considered below.

Effects on the eyes

Using DSE does not damage the eyes or the eyesight, nor does it cause deterioration in existing eye conditions. However, pre-existing eyesight defects may be highlighted by the intense visual effort that such work entails. Frequent changes in the required focal length for the visual task

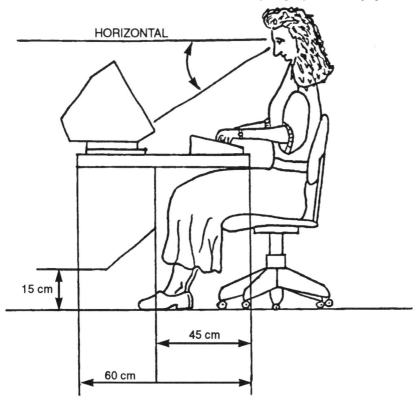

HORIZONTAL

15 cm

45 cm

60 cm

Figure 11.1 The correct position while working at display screen equipment

(from document holder to screen to keyboard) may show up more sluggish eyesight accommodation in older operators, resulting in symptoms of eye fatigue such as sore eyes and headaches. Similarly, eyesight correction may be required to read the screen, which is at a different distance (30 cm) from the normal reading position. Eye fatigue may also result from poor positioning or glare. The requirement for eye and eyesight tests within the regulations addresses these issues.

Fatigue and stress

The relationship between the operator and the display screen is often more intense than that experienced in typewriting. The speed of operation and the lack of natural breaks (such as changing the paper) may be influential here. In many operations there is less operator control of speed of response – the machine may seem to be working the operator. Systems of work should ensure that relaxation periods are inevitable.

Radiation

There is a continually expressed anxiety that display screens emit damaging ionising and non-ionising radiation. Exhaustive tests by the National Radiological Protection Board have shown that emission levels are well below national and international safe limits and there is no risk.

Facial dermatitis

This is another red herring, probably associated with the unfounded belief that operators are being exposed to damaging radiation.

Pregnancy

Anxiety about risks to pregnant operators arose from the concerns over radiation and the apparent increased incidence of pregnancy abnormalities. A number of reliable studies have demonstrated that there is no increased risk of miscarriage or abnormality.

Epilepsy

DSE does not cause epilepsy. However, photosensitive epilepsy occurs in a small percentage of the population. Individuals with this disorder may be susceptible to flickering screens. Such individuals should seek medical advice before beginning work with DSE.

Work-related upper limb disorder

There is no doubt that work with DSE can cause upper limb symptoms. Attempts have been made to group symptoms under various titles such as repetitive strain injury (RSI) and work-related upper limb disorder (WRULD). WRULD is certainly a better term because it does not include the requirement of repetitive action. However, with such a variety of symptoms, these umbrella terms tend to obfuscate rather than clarify the situation. Sitting in a fixed and often unsuitable position for long periods, undertaking repetitive finger, hand or wrist movements, particularly if there are time or other pressures to finish the work, can result in pain and stiffness in the neck, shoulders and arms. Generally, these symptoms disappear with rest, but in certain cases, where symptoms are ignored and pain is 'worked through', more disabling and long-term symptoms may develop and be labelled as some form of WRULD.

Such disorders have been prescribed (entitling the sufferer to worker's compensation) since 1908. Conditions such as telegraphist's cramp and writer's cramp were prescribed between 1908 and 1922. These conditions were described as cramp of the hand or forearm owing to repetitive movements. The 1965 Annual Report of HM Chief Inspector of Factories stated that:

Traumatic inflammation of the tendons or tendon sheaths of the hand or muscles inserted in the lower forearm is a non-infective condition affecting the musculo-tendinous function of the synovial lining of the tendon sheaths or the tendons themselves. It is caused by the constant repetition of small quick movements. The clinical features are local pain, swelling, tenderness and sometimes redness. In the absence of complicating factors, the condition usually subsides with rest. The occupational cramps form a somewhat similar set of conditions which occur in people who have to carry out repeated fine movements such as writers, telegraph operators and the like. An underlying psychoneurosis is suspected in many cases but the condition does not often respond to psychological treatment.

(cited in Tindall 1993)

In October 1990 the Health and Safety Executive published Guidance Note MS10 'Work-related Upper Limb Disorders: A Guide to Prevention'. The term 'upper limb disorder' encompasses a range of conditions affecting the soft tissues of the hand, wrist, arm and shoulder. The guidance note describes symptom patterns, possible causal links, preventive measures and the elimination of risk. Although employer responsibility is clearly established, there is less help with diagnosis, treatment and rehabilitation.

It will be clear that the term WRULD includes a number of conditions, some of which such as tenosynovitis and carpal tunnel syndrome can be diagnosed by objective tests; others may be less amenable to objective evaluation and can be diagnosed on symptomatology alone. Training of operators in the early identification of symptoms is essential. This can result in early intervention both in terms of treatment and correction of unsatisfactory work practices and conditions. There is little doubt that WRULD can, and should, be avoided.

It is clear, therefore, that the only significant health risk from the operation of DSE is WRULD. Every effort should be made to ensure that work organisation does not allow it to develop, and that operators are provided with information on the avoidance and early identification of the condition.

Over the years there have been many cases of claims against employers in relation to symptoms arising from repetitive manual tasks (e.g. *Mountenay v. Bernard Matthews plc* 1993). This has established some basic principles:

- The employee must be warned of the risk.
- The employee must be alerted to the possible symptoms.
- The employee must be trained in the correct posture and use of DSE.
- The employee must be advised to report problems immediately.

A court will consider whether the operation was known to be associated with WRULD, whether the employer attempted to reduce the risk, and whether the symptoms were a result of the work. In a judgment in 1993, the judge ruled that RSI did not exist as an identifiable syndrome, although this decision is likely to be appealed against (*Murghal (Rafiq) v. Reuters* 1993). Stevenson has reviewed the current legal position (Stevenson 1994).

Training

Employers must ensure appropriate DSE health and safety training, which should cover the following issues:

- information on the regulations;
- the user's role in carrying out assessments;
- how to undertake workstation assessments;
- the user's role in the recognition of developing health problems;
- simple explanations of the causes of risk;
- user-initiated actions and procedures to reduce risk;
- organisational arrangements for reporting problems.

Adequate basic training can be carried out in an hour, if there are well-prepared course material, guidance notes and instructions which can be taken away and studied. All training must be recorded.

Eye and eyesight tests

There has been considerable discussion on the rather Draconian requirements of the EC legislation in this context, and it has undoubtedly been fuelled by the self-interested parties. In most organisations it has been possible for employers and employees to reach a reasonably practicable mode of operation, bearing in mind that an employee is given the right to have an appropriate eye and eyesight test on request and that the test is an entitlement not an obligation of the employee.

The tests should be made available to all new DSE users, to all current DSE users following promulgation of the legislation (December 1992), and at intervals as suggested by the examining ophthalmic optician or medical practitioner, and to all DSE users who have developed related eye symptoms.

In practice, most large organisations have offered eyesight screening to all users. This has generally been carried out by in-house occupational health staff using apparatus such as the Master Vision Screener (MAVIS) or the Keystone Vision Screener. A low level of previously unrecognised, uncorrected disorders has been detected (less than 1 per cent in these surveys). Certain opticians have developed facilities and expertise for testing DSE users with particular reference to middle-range vision (50–60cm) and organisations can arrange contracts for eye testing as required. An example of an eye and eyesight test policy is given on page 186. Most employers have found that there has not been an enormous demand for full eye testing.

Conclusion

The use of DSE increases daily in the workplace, thereby potentially putting more employees at risk. Consideration of risk factors at the planning and development stage can avoid long-term problems. (A sample policy is given at the end of this chapter; see pages 187–90.) Contracts with suitable suppliers for chairs and eye tests can help to lower the potential cost of implementing the regulations.

Risk assessment form for VDU Workstation

Name of user . . . .
Department/Service . . . .
Team . . . .
Building . . . .
Room no. . . . .
Name of manager confirming
assessment . . . .
Manager's signature . . . .
Date . . . .

1. Equipment

1a General

Has the VDU been tested in the last 12 months?   Yes   No
If yes, give date

1b Screen/Display

Are the images clear?   Yes   No

Are the images stable?   Yes   No

Is the screen provided with contrast and brightness controls?   Yes   No

Is the screen free from reflections and glare?   Yes   No

Is the angle of tilt adjustable?   Yes   No

Is it possible to use a separate base or adjustable table for the screen?   Yes   No

Is the screen regularly cleaned?   Yes   No

If the answer is no to any of these questions steps should be taken to comply with the regulations.

1c Keyboard

Is the keyboard detachable or movable?   Yes   No

Has it got a shallow keyboard slope (10–20 degrees)?   Yes   No

Is the surface non-reflective, and are the keys well contrasted and legible?   Yes   No

2. Workstation

2a Desk

Is the desk stable?   Yes   No

Is the clearance from the floor to the underside of the desk between 66 and 73 cm for non-adjustable desks and 66 and 77 cm for adjustable desks?   Yes   No

Does it allow for knee clearance?   Yes   No

Is there space to stretch the legs while sitting at the desk, e.g. 45 cm from the front of the desk to 60 cm foot room?   Yes   No

Is the desk deep
enough to allow for
flexible arrangement
of equipment and
documents (minimum
60 cm, optimum
80 cm)?   Yes   No

Is the desk long
enough to allow for
flexible arrangement
of equipment and
documents (minimum
120 cm, optimum
160 cm)?   Yes   No

Is there enough support
for hands and wrists
(approx. 5–10 cm in
front of the keyboard)?   Yes   No

Is the desk free from
sharp edges that can
cut into the wrists?   Yes   No

2b Chair

Does the chair
provide good support
for the back and
buttocks?   Yes   No

Does it have a five-star
base configuration on
castors?   Yes   No

Is it adjustable in
height?   Yes   No

Does the back rest
adjust in height and tilt? Yes   No

Does it swivel to give
access to work surface
and storage?   Yes   No

2c Printer

Is the printer
satisfactorily sited in
relation to:

– accessibility?   Yes   No

– the proximity of
  other workers?   Yes   No

2d Document holder

Is a document holder
necessary?   Yes   No

If yes, is there one
available and is it
adjustable in height
and angle; has it a
matt surface?   Yes   No

2e Working posture

Is the distance
between the screen
and the operator's
forehead approximately
35–70 cm?   Yes   No

Are the operator's
eyes level with the top
of the screen?   Yes   No

Are the operator's
hands and forearms
at an angle of
approximately 90° to
the body?   Yes   No

When the operator sits
back in the chair, is
there a 90° angle in
hips and knees?   Yes   No

If the operator's feet
cannot touch the floor,
is a foot rest provided?   Yes   No

3. Environment

3a Layout

Is the space in the work area as a whole:

- sufficient to allow
  mobility?   Yes   No

- relevant to the
  type of work (e.g.
  telephone usage,
  level of concentration,
  dealing with the
  public)?   Yes   No

- capable of allowing
  easy escape in the
  event of fire?   Yes   No

- sufficient for the
  number of people
  and amount of
  furniture and
  equipment?   Yes   No

Does the space at the desk/work-station:

- accommodate the
  amount of equipment
  used?   Yes   No

- accommodate the
  work undertaken?   Yes   No

- accommodate
  manuals, files, etc.?   Yes   No

- allow for a
  comfortable working
  posture?   Yes   No

Are all electrical cables
masked or ducted?   Yes   No

3b Lighting

Is the lighting in the
workplace and at the
desk suitable and
efficient?   Yes   No

3c Heating and ventilation

Is there an adequate
heating and ventilation
system?   Yes   No

Is the temperature
comfortable?   Yes   No

Is the relative humidity
comfortable?   Yes   No

3d Noise

Is the equipment
sited satisfactorily so
that noise is not a
nuisance?   Yes   No

Is the background
noise level low
enough to work
comfortably?   Yes   No

4. Task design and software

Is the keyboard
work regularly
interrupted by other
activities away from
the VDU workstation?   Yes   No

Is the software suitable
for the task?   Yes   No

Is the software easy
to use and adaptable?   Yes   No

5. Health

Has the user encountered
any health problems?   Yes   No

6. Remedial action for risk reduction in order of priority and time frame

	Action	Time frame	Person responsible
1.	. . . . .	. . . . .	. . . .
2.	. . . . .	. . . . .	. . . .

3.	. . . . .	. . . . .	. . . .
4.	. . . . .	. . . . .	. . . .
5.	. . . . .	. . . . .	. . . .
6.	. . . . .	. . . . .	. . . .
7.	. . . . .	. . . . .	. . . .
8.	. . . . .	. . . . .	. . . .
9.	. . . . .	. . . . .	. . . .

7. General comment

Date by which action is to be taken:

Date of reassessment:

A chair-purchasing policy

1. Introduction

The Display Screen Equipment (DSE) Regulations came into effect on 1 January 1993. The aim of the regulations is to reduce the risk to staff of an occupational injury resulting from continuous use of DSE. They provide detailed information on minimum health and safety standards for workstations, including information on the characteristics of a suitable chair.

2. Time frame

All new workstations must comply fully with the legislation immediately. In addition, if a health problem is identified existing workstations must be upgraded immediately. All other workstations must comply by 31 December 1996.

3. Minimum requirements for chairs used at workstations

(a) The chair must be stable and allow the user easy freedom of movement and a comfortable position.
(b) The seat must be adjustable in height.
(c) The seat back must be adjustable in both height and tilt.
(d) The chair must have a five-star base configuration on castors or gliders. Please note that gliders must be requested for use in areas with hard floors as castors can be dangerous in these situations.

4. The recommended chairs

The [name] series of chairs should be ordered when replacing chairs used at DSE workstations because:

(i) they meet all the specifications of the DSE Regulations 1992;
(ii) they offer variety to the user including:

 - a choice of three back rests, including a back with lumbar support
 - a wide choice of colours and three types of material
 - a choice of castors or gliders
 - availability with or without arm rests;

(iii) they comply with the fire regulations;
(iv) they are competitively priced and good value for money.

5. Ordering of chairs

To facilitate correct ordering, the supplies department will keep two sample chairs on each site. This will include a chair with a medium back support and a chair with lumbar support. Samples of materials and colours are also available through the supplies department.

On receipt of an order for a chair, the supplies department will only order [name] chairs.

In exceptional circumstances a different type of chair may be ordered but only with the approval of the occupational health department.

A policy on eye tests for users of display screen equipment

Introduction

The Health and Safety (Display Screen Equipment) Regulations came into force on 1 January 1993.

Regulation 5 requires employers to provide users with an appropriate eye and eyesight test. This policy outlines the arrangements for eye and eyesight testing in this organisation.

The policy

1. Staff identified as users of display screen equipment, as defined by the above regulations, will be offered an eyesight test by the occupational health nurse using the Keystone vision screening equipment.

2. Staff found to have difficulty with vision at the distance used in display screen equipment work or those who choose initially to have a full eye examination will be referred to [name] opticians, under the VDU Eyecare Plan.

3. Staff may go to any [name] branch of their choice, but referrals will only be accepted on production of an appropriate referral letter from the occupational health department.

4. The eye examination and any subsequent glasses required for VDU work only will be paid for by the company [cost as agreed]. If bifocal or other more expensive lenses or frames are chosen, any additional cost will be at the employee's expense.

5. Staff found to have eye problems other than those related to middle distance vision will be liable for all costs incurred other than the initial eye test.

6. Repeat eye tests are usually required approximately every two years. More frequent eye tests will only be paid for by the organisation if prior arrangements are made with the occupational health department and on production of the appropriate referral letter.

Manager's responsibility

1. It is the manager's responsibility to identify 'users' in each department and inform them of this policy.

2. Users can make appointments for Keystone vision screening by contacting the occupational health department.

A policy on the implementation of the display screen equipment regulations 1992

1. Introduction

Display screen equipment (DSE) is the term used to describe the electronic display equipment that forms part of a computer system. Until now the safe use of DSE has been governed by the general provisions of the Health and Safety at Work etc. Act 1974, but new, specific legislation – the Health and Safety (DSE) Regulations 1992 – came into force on 1 January 1993.

The aim of the regulations is to reduce the risk to staff of an occupational injury resulting from the continuous use of DSE. Non-compliance with the regulations is a prosecutable offence.

The regulations refer to workstations and users throughout. A workstation includes the desk, chair, DSE and the general space surrounding this assembly including the lighting. A user is defined in Appendix 1.

2. Risk assessment (Regulation 2)

2.1 Undertaking risk assessment

The DSE regulations require employers to carry out an analysis of all workstations to assess potential risks to users, and identify ways of reducing these risks.

The assessment must take into account all aspects of the task, the workstation, working environment and any individual factors relating to the operator. There must be a further assessment following any changes to the working environment.

Guidelines on undertaking the risk assessment, together with a risk assessment form, are attached.* The guidelines outline the minimum standards required, and give practical advice on how to overcome some common problems.

Assessments are the responsibility of heads of department and should be carried out by them or their nominated representatives. Training in undertaking risk assessment is provided by the occupational physio-therapist through the occupational health department.

---

* Please note that guidelines are not attached. This is a sample policy only.

2.2   Time frame (Regulation 3)

Any workstation brought into use after 1 January 1993 must meet the minimum requirements. All other workstations must meet the minimum standards by 31 December 1996. However, where a member of staff experiences symptoms as described in 6.2 below, remedial action should be taken without delay.

2.3   Daily work routine of users (Regulation 4)

The regulations require employers to plan users' activities so that there are periodic breaks or changes of activity, reducing the time spent continuously at the keyboard. A guide would be a five-minute break from the screen every hour.

3.   Eyesight tests (Regulation 5)

The regulations require employers to ensure that users are provided with an appropriate eyesight test on commencing work with the DSE and on request.

The organisation's policy on eye tests is attached.*

4.   Provision of training (Regulation 6)

Employers must provide health and safety training for all new employees who will be users of DSE and for all current users as soon as possible. Whenever arrangements, including software at their work-stations, are modified, further training must be provided.

5.   Information

Managers must ensure that users are informed about the action being taken to meet the regulations, their entitlement to eye and eyesight tests, and the training arrangements.

6.   Health

6.1   Some operators may experience symptoms of eyestrain where their vision is inadequate without correction for work with DSE. They should be referred to the occupational health department immediately.

---

* Please note that the policy on eye tests is not attached. This is a sample policy only.

6.2 Any operator who experiences discomfort in his or her arms, wrists or hands, back, shoulders or neck must be referred to the occupational health department for advice immediately.

6.3 To reduce fatigue, DSE-based tasks should be organised in such a way that spells of concentrated work at the terminal are spaced throughout the day. Staff should exercise during breaks to relieve muscle tension which may be brought about by continuous work in one position.

7.   Manager's responsibilities

7.1 It is the manager's responsibility to ensure that a risk assessment is carried out on all workstations in his or her department. A written assessment must be kept and reviewed when changes are made.

7.2 Managers must prioritise action on any risks identified and take steps to reduce the risks as quickly as possible. All workstations must comply fully with the regulations by 31 December 1996.

7.3 Where a health problem is identified (as in 6.2 above) the workstation must be upgraded to the minimum standard without delay.

7.4 Managers must advise staff of their entitlement to eye and eyesight tests.

7.5 Managers must teach (or arrange teaching for) all users of DSE how to set up and adjust their workstations to suit their individual needs.

Appendix I

The definition of a user

A person is defined as a user if he or she meets one or more of the following criteria:

(a) The individual depends on the use of DSE to do the job, as alternative means are not readily available for achieving the same result.
(b) The individual has no discretion as to use or non-use of the DSE.
(c) The individual needs significant training and/or particular skills in the use of DSE to do the job.
(d) The individual normally uses DSE for continuous spells of an hour or more at a time.
(e) The individual uses DSE in this way more or less daily.
(f) Fast transfer of information between the user and the screen is an important requirement of the job.

(g) The performance requirements of the system demand high levels of attention and concentration by the user; for example, where the consequences of error may be critical.

Chapter 12
Manual handling of loads

Thus to persist
In doing wrong extenuates not wrong,
But makes it much more heavy.
          (William Shakespeare, *Troilus and Cressida*, II. ii. 186)

More than a quarter of reportable accidents are associated with manual handling and there is evidence of increasing litigation with substantial damages being awarded. The Health and Safety at Work Act requires all employers to ensure the health and safety of their workforce. Special regulations now apply to manual handling. Under these regulations employers are required to assess all manual handling tasks. This chapter gives guidance on the preliminary and full assessment of these tasks using a checklist. The type of task, the nature of the load, the working environment and individual capabilities are considered. There is a simple guide to the reduction of risk and an outline of training and recording requirements.

Introduction

Low back pain occurs at some time in 80 per cent of the population. It is not just a problem for industries where there is a heavy manual handling component. Each year more than a quarter of reportable accidents are associated with manual handling. Sixty-seven per cent of these cases are described as suffering from back sprain or strain (see Figure 12.1).

These injuries result from incorrect handling procedures, repetition of potentially dangerous operations, and handling unpredictable loads. The cause may therefore be incidental or cumulative. The incidence of back pain in non-manual workers is similar to that of manual workers (Anderrson 1979). However, the degree and length of incapacity are much greater in the latter group. The underlying pathology/disease is often unclear and it is fortunate that in four out of five employees the condition

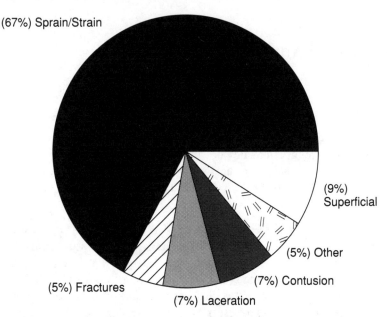

(67%) Sprain/Strain

(9%) Superficial

(5%) Other

(7%) Contusion

(7%) Laceration

(5%) Fractures

Figure 12.1 Types of injury

is self-limiting and is resolved within three weeks (Benn and Wood 1975). This is particularly fortunate as it appears that treatment is more likely to be palliative than curative. Nevertheless, it has been estimated that in nurses alone back symptoms account for 764,000 lost working days annually (Stubbs *et al.* 1983).

Legislation

The Health and Safety at Work etc. Act 1974 places a general duty on employers to ensure the health and safety of the workforce. This is underpinned by industry or hazard specific legislation such as the Control of Substances Hazardous to Health (COSHH) Regulations. At the beginning of 1993 specific regulations in relation to manual handling came into force. The Manual Handling Regulations 1992 form part of a group of six sets of regulations (the so-called six pack) enacted in line with the EC Framework Directive:

- Management of Health and Safety at Work Regulations;
- Health and Safety (Display Screen Equipment) Regulations;
- Manual Handling Operations Regulations;
- Provision and Use of Work Equipment Regulations;

- The Personal Protective Equipment at Work Regulations;
- Workplace (Health, Safety and Welfare) Regulations.

The Management of Health and Safety at Work Regulations also impose a general duty on employers to assess and manage risks in the workplace. The aim of the Manual Handling Regulations is to reduce the risk to staff from any operation requiring manual handling as far as is reasonably practicable.

Employers are required to follow a clear hierarchy of measures:

- To avoid hazardous manual handling operations as far as is reasonably practicable.
- To make a suitable and sufficient assessment of any hazardous manual handling operation which cannot be avoided.
- To reduce the risk of injury from these operations as far as is reasonably practicable.

The well-established phrase 'as far as is reasonably practicable' gives employers some leeway in the measures that they need to take. However, there can be no clear rules on what is reasonably practicable. It is also probably important to note here that this phrase does not appear in the Framework Directive.

Assessment

Preliminary assessment

It is obvious that a full assessment of every manual handling task could be a major undertaking and the Health and Safety Executive has, therefore, issued a guidance note on preliminary assessments which will determine where full assessments are required. Guidance notes do not have the force of law but can be taken into consideration by inspectors when checking compliance. They suggest considering four major activities:

- lifting and lowering loads;
- carrying loads;
- pushing and pulling;
- handling while seated.

Lifting and lowering loads
The weight of the load to be lifted is only one part of the assessment. For these operations the assessment should also take into account the position

of the arms. The weight that can be lifted safely diminishes if the arms are outstretched or if the movement goes above shoulder height. The size and shape of the operator are clearly of some significance in this calculation. Where the arc of movement is considerable, a full assessment will be required. If twisting is necessary or the movements have to be repeated more than 30 times per hour, the safe load size will be reduced and full assessment is required. Suggested load limits are shown in Figure 12.2.

Carrying loads
The basic weight guidelines for lifting and lowering also apply here. However, they apply only when the load is carried close to the body and

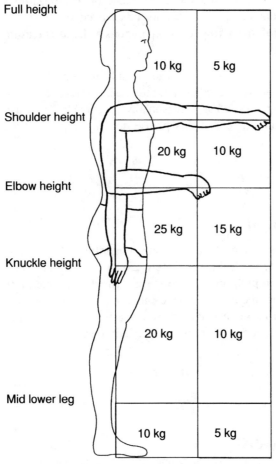

Figure 12.2 Suggested load limits

for distances of up to ten metres. If this is not the case, full assessment is required.

Pushing and pulling
It is suggested that a 15 kg force would be appropriate for pulling loads on wheels if the hands are between knuckle and shoulder height. If these criteria are not met, full assessment is required. Unfortunately, tests for estimating the force of such movements are not readily available and most employers are trying to develop practical tests.

Handling while seated
This is not a position of strength and if there is more than a 5 kg load, or if considerable twisting is involved, full assessment will be required.

Full assessment

The first question to ask is whether the manual handling part of the operation can be eliminated completely; for example, by carrying out certain tasks *in situ* or using some form of mechanisation. If the task cannot be avoided and the preliminary assessment shows that there is a potential risk, the employer must proceed to a full assessment. The individual undertaking the assessment should understand:

- the requirements of the regulations;
- the nature of the task;
- individual capabilities;
- how to identify high risk activities;
- how to reduce risk.

For the majority of manual handling tasks the well-informed manager should be able to carry out the assessment with the help of others within the organisation, such as occupational health or safety professionals. Detailed checklists are particularly helpful to the average manager in making the assessments. An example of a checklist is shown on pages 201–4.

Certain complex operations may require greater expertise and this can be provided by external consultants. (See Useful Addresses, pages 211–15.)

Where there are many different and varied manual handling tasks, it may be possible to group these so that generic assessments can be made.

The full assessment should be made under five categories:

- the task;
- the load;
- the working environment;
- individual capability;
- personal protective equipment.

The task

Several factors are known to increase stress on the lower back. These include:

- holding the load at a distance from the trunk;
- not being able to stand squarely with the feet flat on the ground;
- twisting round while holding the load and reaching upwards;
- the distance through which the load has to be lifted;
- the distance over which the load is carried;
- pushing or pulling with the load below knuckle height or above shoulder height;
- the frequency with which the task has to be undertaken;
- the possibility of rest periods.

Handling required while seated may be particularly risky as this position precludes the use of the leg muscles.

The load

The actual weight of the load is only part of the potential risk. Traditional guidelines on acceptable weights have not taken into account all the other factors involved. These include whether the load is bulky or unwieldy; too big to allow a suitable grip or good vision; or unstable with an eccentric centre of gravity. All these factors will increase the risk to the handler.

Similarly, if the load is difficult to grasp because it is smooth or slippery, greater effort is likely to be required. Loads which are unstable because of shifting contents may produce sudden stresses, thereby increasing the risk to the handler. Handling people or animals creates additional risk, because of the unpredictability factor, the lack of rigidity and the handler's desire not to cause injury. Other types of load may have sharp corners or rough surfaces which may discourage or prevent good gripping and cause other injuries.

The working environment

An unsatisfactory working environment may considerably increase the risk from manual handling activities. The usual problem is inadequate space, hindering a good ergonomic approach to the task and involving additional twisting and manoeuvring. Slippery or uneven floors may create further risk.

Individual capability
The concept of individual capability is relatively new in general health and safety legislation and may need to be considered carefully as a general principle. It has the potential for excluding those who have a disability, or who are less fit, from many places of work. It should also be remembered that medical screening of asymptomatic employees has not been shown to reduce the risk of back injury.

The assessment of individual capability is in three main areas: the requirements of the task, the individual's state of health, and the individual's previous training. In general, the lifting strength of women is less than that of men but, of course, there is a wide overlap. Similarly, although physical capability varies with age and the risk of manual handling injury appears to be greater for employees in their teens or over 50, there is, as with sex, a considerable range of capabilities. If unusual strength or height is required, the risk should in general be regarded as unacceptable. Employers are required to make allowances for pregnancy. Pregnancy may increase the risk of injury because of hormonal changes affecting the musculo-skeletal system and obvious postural problems in the later stages.

The employer is also required to take into account any known health problem which may affect the individual's ability to undertake manual handling tasks. This could include previous or current back or other musculo-skeletal problems and cardiac or respiratory disease. Where there is doubt, it is sensible to seek medical advice. When seeking that advice, it is important to give a clear description of the job requirements. This is particularly so where there is no occupational health service – general medical practitioners may have only vague ideas about the actual task. It should not be left to the employee to describe the job to the doctor since this is fraught with possibilities of misinterpretation. Finally, the employer must consider whether the individual has received adequate training for the task.

Personal protective equipment
Certain manual handling work may be associated with other potential health risks which require the use of personal protective equipment, such as gloves or breathing apparatus. Any such equipment will increase the problems of the task and should form part of the risk assessment.

Reducing the risk of injury

Having carried out a full assessment, the employer should have a hierarchy of risk for the various manual handling tasks which will enable prioritisation of action required from high to low risk. Where significant risk has been identified, it is the responsibility of the employer to take appropriate steps to reduce the risk of injury to the lowest level reasonably practicable. To comply with the regulations, where there is immediate risk of injury, recommendations must be implemented without delay. Where risk of injury is not immediate, there should be a planned implementation period. The approach to risk reduction will depend on the nature of the task and the circumstances. Where the operation is relatively unchanging, improvement may be more appropriately brought about by changing the nature of the task. In other circumstances it may be appropriate to improve the handling techniques.

Task and workplace design

Changes to the layout of the task may take the form of:

- altering the sequence of the operation to reduce the need for twisting and bending;
- introducing team handling;
- providing handling aids.

The risk of manual handling can be reduced by alterations in the work routine. For example, the need to maintain a fixed position while supporting the load should be reduced, time for recovery built into the system of work, and where possible repetitive tasks paced by the individual. Rotating tasks during the working day may allow one group of muscles to recover while a different task is addressed.

Handling aids do not remove the manual handling element of the task but they allow the more efficient use of bodily forces. Hoists, trolleys and chutes are obvious ways of reducing the manual lifting component. Similarly, devices such as hand-hold hooks and suction pads can increase safety by facilitating handling. Where team handling is introduced, there should be enough space for each handler to grip the load. One of the team must be in charge and team members should have similar physical capabilities.

If some form of mechanisation is introduced, regular maintenance is essential. Trolleys with sticking wheels may in themselves provide a significant health risk because of erratic resistance and movement.

The importance of developing written safe working procedures cannot be overestimated.

The load

Various adjustments to the load can be made in order to reduce the risk of injury. For example, it may be possible to break down the load into smaller packages in size and weight. Where the size or surface texture of the load makes it difficult to grasp, handles or grips may help (hand-holds should be as wide as the palm and deep enough to accommodate knuckles). It may be possible to place the load in an easily handled container or a sling. In some cases hot or cold loads can be placed in suitably insulated containers. This is clearly not an option in canteen work where size should enable easy handling. When the surface of the load presents a cutting hazard which cannot be removed, protective gloves will be necessary.

The environment

Increasing the size of doors or corridors may not be practical. However, considerable improvement can often be achieved by better housekeeping, such as removing obstructions from the work area. Where uneven floor or ground surfaces are largely unavoidable, as in outdoor work, every effort should be made to provide firm ground or suitable coverings in the work area. Where this is impossible, the size of load which can safely be handled will be considerably reduced. Similarly, outdoor tasks will be performed at greater risk when the operator is cold or where weather conditions impinge on the activity.

Individual capability

The provision of different or modified tasks for pregnant workers and those with significant health problems may be necessary and should be considered on an individual basis.

Information and training

All employees engaged in manual handling operations, particularly those which have required full assessment, should receive information and training. This should not be seen as an alternative to reducing the risk of injury by modifying the operation as described above, but as a complement to it. Task-oriented training has an important part to play in risk reduction. The programme must include practical and theoretical training and information on the following areas:

- good handling techniques;
- how to recognise potentially hazardous operations;
- the proper use of handling aids;
- the importance of the working environment;
- factors affecting individual capability.

Record keeping

Records must be kept of each assessment and these should be available to employees. Records must also be kept of each employee's training in safe manual handling.

Continuing activities

Assessments should be reviewed if there is any change in the operation or if there has been a reportable injury. Employers also have a duty to monitor the use of safe working practices and check regularly that these practices are effective.

Written safe working procedures should be clearly displayed and available to employees at the appropriate sites of work. From time to time employees' knowledge of these procedures should be checked.

Conclusion

The requirements of the Manual Handling of Loads Regulations have been seen by many employers as equivalent to, if not more Draconian than, those of the COSSH Regulations. However, the Health and Safety Executive has advocated a practical and realistic approach, and implementation should not be too daunting if the employer deals with it in an orderly fashion. The steps to be taken involve:

- a list of all manual handling tasks;
- a preliminary assessment to identify those tasks which need full assessment;
- a full assessment of those tasks involving potential risk;
- risk reduction;
- training;
- monitoring;
- record-keeping.

A risk assessment checklist

The load

(a) Weight.
(b) Shape/size: Is it too large to be easily handled? Does the shape of the load make handling difficult? While handling the load, is vision unobstructed?
(c) Handling characteristics: Is it difficult to grasp? Is it large/round/ smooth/wet/greasy, etc.?
(d) Stability: Is it unbalanced, unstable, or does it have contents which are likely to shift? Does it lack rigidity? Will the centre of gravity be displaced on lifting?
(e) Danger: Is the load sharp, hot or otherwise potentially damaging?

The task

(a) Is the load held at a distance from the trunk, increasing the level of stress on the lower back or making the load difficult to control?
(b) To perform the operation, does the employee have to adopt an unusual posture, e.g. reaching above the head, twisting, stooping or any combination of these, resulting in increased stress on the lower back?
(c) Does the task involve excessive lifting and lowering distances, e.g. from the floor to above head height, perhaps resulting in the need for a change in grip partway through the operation and increasing the risk of injury?
(d) Does the task involve excessive carrying distances resulting in physical stress and increased risk of injury?
(e) Does the task involve excessive pushing or pulling of the load?
(f) Is there a risk of sudden movement of the load resulting in an unpredictable stress on the body, e.g. freeing a jammed wheelbrake?
(g) Does the task involve frequent or prolonged physical effort, possibly resulting in fatigue?
(h) Is there a sufficient rest and recovery period before the task is repeated, reducing fatigue?
(i) Is the operation performed while seated, preventing the use of leg muscles and perhaps resulting in the operator reaching and leaning forward, thus increasing stress on the lower back?
(j) Is the operation performed by teams of employees, during team operations? Is the load distributed as evenly as possible?

The person

Manual handling injuries are more often associated with the nature of the operations than with variations in individual capability; however, the following should be considered in the risk assessment:

(a) Is unusual strength or height required?
(b) Is specialist knowledge or training required?
(c) Are those who have special health problems at risk?
(d) Are pregnant women at risk?

Lifting aids

(a) Are lifting aids in use?
(b) Are there written instructions, and is training provided for staff who are required to use mechanical devices or handling aids?
(c) Are mechanical devices or handling aids adequately tested and guaranteed by the manufacturer to ensure compliance with the relevant regulations in place at the time?
(d) Are mechanical devices or handling aids tested annually to ensure that no faults have developed and that specifications are still in place?
(e) Are handles provided on equipment used for pushing or pulling articles? Are these adjustable to allow for height differences in staff using the equipment?

Personal protective equipment

(a) Is personal protective equipment provided where there is a specific risk from a load, e.g. gloves, overalls, safety shoes?
(b) Where the load is dusty or fibrous, are goggles, masks and respiratory protection provided?

The Workstation/environment

(a) Is the area well lit?
(b) Are all gangways wide enough to accommodate personnel and equipment with adequate room to manoeuvre, clean and clear from obstruction?
(c) Is the area laid out to minimise the amount of manual effort, twisting, bending, stretching, carrying distance and discomfort?
(d) Is there adequate headroom, where practical, to allow the implementation of good lifting techniques?

A risk reduction checklist

The load

(a) Can the load be made lighter and less bulky, e.g. packaged in smaller containers?
(b) Can the load be made easier to grasp by provision of handles or hand grips, or can it be placed in a container which is itself easier to grasp?
(c) Can the load be made more stable with contents which are less likely to shift?
(d) Can the surface of the load be made less damaging by insulation or a reduction in sharp or rough surfaces on the container?
(e) Can team lifting be introduced for heavy loads?

The task

(a) Can the task layout be improved, e.g. by more ergonomic consideration when designing storage facilities, with heavy loads stored at waist height? Is it possible to reduce the need for excessive carrying distances by the provision of trolleys, etc.? Is it possible to reduce lifting distance by provision of additional storage?
(b) Can the body be used more effectively with loads held closer to the body, elimination of obstruction at floor level when handling loads, replacing lifting with pushing or pulling? Is there secure footing when pushing or pulling?
(c) Can the work routine be improved, minimising the need for fixed postures, reducing handling frequency, rotating staff, provision of rest breaks?
(d) Can handling while seated be reduced, especially any task involving the lifting of a load from the floor while seated? Is the seat provided stable?
(e) Should the task be performed with suitable handling aids, reducing the need for human effort?
(f) Should protective equipment be provided for staff?

The person

(a) Have staff been trained in relation to the task?
(b) Have staff been informed, where possible, of the weight of the load and the risks associated with handling operations?
(c) Have staff with a history of significant health problems been assessed medically prior to employment?

(d) Are staff who have had sickness absence owing to manual handling incidents referred for medical assessment before returning to work?

The workstation/environment

(a) Has removal of space constraints been considered or carried out, e.g. widening gangways, increasing headroom?
(b) Is there a possibility of improving workstation layout, e.g. providing additional shelving, removing obstructions, allowing work to be performed at a reasonable height?
(c) Is it possible to improve the lighting?
(d) Is it possible to improve the flooring, removing slopes or trip hazards?
(e) Is it possible to improve the heating/ventilation to maintain a comfortable thermal environment?

Maintenance

Are maintenance/inspection programmes for all handling aids and protective equipment in place?

Safe working procedures

Are written safe working procedures available for all complex tasks?

Education and training

(a) Have all existing staff involved in manual handling completed an initial, documented, manual handling training programme?
(b) Are arrangements in place for all new staff to complete manual handling training on commencing employment?

Monitoring

Are procedures in place for regular monitoring?

Postscript

The development of occupational health in the UK differs from that in other countries. In the rest of Western Europe occupational health provision is largely statutory, the number of occupational health physicians being determined by staff numbers in a prescriptive manner. In the USA the development of occupational health is related to the cost of sickness absence and company insurance obligations. Insurance premiums can be substantially reduced by putting an occupational health service in place.

In this country we do not have these motivations and development has been either related to health and safety legislation or a genuine attempt to protect and promote the health of the workforce, sometimes philanthropically and sometimes as a 'good business' initiative. This has resulted in an uneven distribution of facilities, which is particularly obvious in large occupational groups such as the National Health Service. Individual hospitals and trusts have vastly different occupational health provision. Although an increasing number are consultant-led, there are still many hospitals where there is not even a trained occupational health nurse. It is surprising that, even when the NHS was 'a whole', there was no support from the Department of Health for the setting of even minimal standards.

As we have seen, an increasing number of organisations are seeking some occupational health input to meet health and safety legislation requirements and in line with *The Health of the Nation* proposals. However, the trend is away from internal services towards the use of independent occupational health consultancies. This is a double-edged sword. It has the potential to improve the quality of occupational health input, as such consultancies are likely to be led by qualified occupational health physicians or nurses. But, since the mode is that of purchaser/provider, it may prove difficult to establish good occupational health practices. For example, although pre-employment medical examination is generally a waste of time, many organisations want it and are prepared to pay for it. External providers may find it difficult to move an organisation towards a more proactive model of health care, with the emphasis on the prevention of ill health rather than individual problem-solving.

It has never been easy to demonstrate cost benefits in occupational health, largely because the results are rarely immediate and, in any case, difficult to calculate. Research is difficult because one is often dealing with an unmatched volunteer population. Probably the most easily demonstrated benefit is related to occupational health involvement in the management of sickness absence.

Unfortunately, this is not a role that occupational health has wanted to highlight as it may be perceived as a tool of management. In this, as in any other aspect of occupational health, the physician has to take a balanced view and give scientifically based opinions favouring neither patient nor employer.

The successful occupational health physician hopes to be described at various times as either the 'tool of management' or 'in the pocket of the unions'. This does not mean that the physician is unable to give clear-cut advice, only that advice may not always be palatable, although it must be sound. The authors do not wish to suggest that such a balancing act is easy. Those who buy into occupational health should expect to receive well-balanced opinions and scientific comment. We hope that this book will have gone some way to demonstrate the value of such specialist advice.

Bibliography

Anderrson, G. B. J. (1979) 'Low back pain in industry: epidemiological aspects', *Scandinavian Journal of Rehabilitation Medicine* II.

Anon. (1975) *Bulletin of the Society for the Social History of Medicine* No. 16.

Audit Commission (1993) *Get Well Soon. A Re-appraisal of Sickness Absence in London.* London: HMSO Publication.

Belbin, R. M. (1981) *Management Teams: Why They Succeed or Fail.* Oxford: Butterworth-Heinemann.

Benn, R. T. and Wood, P. N. H. (1975) 'Pain in the back: an attempt to estimate the size of the problem', *Rheumatism Rehabilitation* 14.

Bertera, R. L. (1991) 'The effects of behavioural risks on absenteeism and health care costs', *Journal of Occupational Medicine* 33 (11) November.

Blick Time Systems Study (1993) *Personnel Today* May: 48.

British College of Optometrists (1993) *Work with Display Screen Equipment – A Statement of Good Practice.*

British Standards Institute (1992) *Ergonomics of Design and the Use of Visual Display Terminals in Offices* Parts 5 and 6.

Chartered Institute of Building Services Engineers (1989) *Lighting Guide: Areas for Visual Display Units.*

Clothier, C. M. (1994) *The Allitt Inquiry.* London: HMSO.

Communicable Diseases Report (CDR) (1987) *AIDS and Employment* No. 15 September/October.

Communicable Diseases Report (CDR) (1993) *The Incidence and Prevalence of AIDS* Vol. 3, Supplement 1.

Confederation of British Industry (CBI) (1993) Percom Survey: *Too Much Time Out.* London: CBI.

Confederation of British Industry (CBI) (1993) *Working for Your Health.* London: CBI.

Cooper, C. (1988) *Occupational Stress Indicator.* Windsor: NFER.

Day, N. (1993) 'Patterns of heterosexual HIV infection', *Journal of Royal Society of Medicine.*

Department of Health (1987) *AIDS Information for the Work Place.* London: HMSO.

Department of Health (1993) *The Health of the Nation.* London: HMSO.

Edwards, F. C., McCallum, R. I. and Taylor, P. J. (eds) (1988) *Fitness for Work: The Medical Aspect.* Oxford: Oxford University Press.

Employers' Forum on Disability (1992) *Valuing Ability.*

Employment Department Group (1993) *Code of Good Practice on the Employment of Disabled People.* London: HMSO.

Employment Department Group (1994) *Labour Force Surveys 1984 and 1993.* London: HMSO.

Employment Medical Advisory Service (EMAS) (1977) *Occupational Health Services – The Way Ahead.* London: HMSO.

Faculty of Occupational Medicine (1993) *Guidance on Ethics for Occupational Physicians.*

Gabel, H. D. and Colley Niemeyer, B. (1990) 'Smoking in a public health agency: its relationship to sick leave and other life-style behaviour', *Southern Medicine Journal* 1.

Handy, C. B. (1976) *Understanding Organisations*, 4th edn. Harmondsworth: Penguin Business Library.

Health and Safety Executive (1972) *The Manual Handling of Loads Regulations.* London: HMSO.

Health and Safety Executive (1973) *The Health and Safety at Work etc. Act.* London: HMSO.

Health and Safety Executive (1990) *Work Related Upper Limb Disorder: A Guide to Prevention.* London: HMSO.

Health and Safety Executive (1992) *Display Screen Equipment Work: Guidance and Regulations.* London: HMSO.

Health and Safety Executive (1992) *Visual Display Units.* London: HMSO.

Health and Safety Executive (1992) *The Manual Handling Guidance on Regulations.* London: HMSO.

Health and Safety Executive Research Report (1993) *Occupational Health Provision at Work.* London: HMSO.

Health and Safety Information Bulletin (1991) *AIDS and the Work Place 1* No. 186.

Health and Safety Information Bulletin (1991) *AIDS and the Work Place 2* No. 187.

Hemminki, K. (1985) 'Spontaneous abortions', *Journal of Epidemiological Community Medicine* 39: 141–7.

Herzberg, F. (1966) *Work and the Nature of Man.* New York: World Publishing Co.

Holmes, T. M. and Rahe, (1967) 'The Social Readjustment Scale', *Journal of Psychosomatic Medicine* II.

Institute of Personnel and Development (1993) *The Business Case for Family Friendly Provision.* London: IPM.

Institute of Personnel and Development (1993) *Corporate Culture and Caring.* London: IPM.

Jenkins, R., Harvey, S., Butler, T. and Thomas, R. L. (1992) 'A six-year longitudinal study of the occupational consequences of drinking over the safe limit of alcohol', *British Journal of Industrial Medicine* 49(5) May.

Kinnersley, P. (1973) *The Hazards of Work: How to Fight Them.* Love and Malcolmson Ltd.

Lazarus, R. S. (1966) *Psychological Stress and the Coping Process.* New York: McGraw-Hill.

Lazarus, R. S. (1971) 'The concept of stress and disease', in L. Levi (ed.) *Society Stress and Disease* Volume 1. Oxford: Oxford University Press.

Levinson, H. (1962) *Men, Management and Mental Health.* Oxford: Oxford University Press.

Leviton, L. C. (1989) 'Can organisations benefit from worksite health promotion?', *Health Service Research*, University of Michigan, 24(2) June.

Lifeshield Foundation (1990)

Maslow, A. (1970) *Motivation and Personality.* New York: Harper and Row.

O'Sullivan, J. S. (1992) unpublished data.

Peters, T. (1985) *Not Just Another Publishing Company.*

Personnel Today (1992a) 'Employees in manufacturing and non manufacturing 1980–1992, September: 48.

Personnel Today (1992b) 'Closed shops 1980–1990', October: 48.

Personnel Today (1993a) 'Trade union membership', April: 48.

Personnel Today (1993b) 'Working patterns', May: 48.

Personnel Today (1994) 'Strike rates 1983–1992', January: 56.

The Tom Peters Group – Training material – *Implementing in Search of Excellence.*

Plsek, P. E. and Onnias, A. (1989) edited by John F. Early for Juran Institute Inc. – training material/flow diagrams.

Porter, L. W. and Lawler, E. E. (1968) 'What job attitudes tell about motivation', *Harvard Business Review* January–February.

Raffle, F. A. B., Lee, W. R., McCallum, R. I. and Murray, R. (eds) (1991) *Diseases of Occupations.* London: Edward Arnold.

Reddy, M. (1992) 'Counselling: its value to business', in *Prevention of Mental Ill Health at Work.* Conference report. London: HMSO.

Royal College of Physicians (1987) 'Medical consequences of alcohol abuse: a great and growing evil'.

Ryan, J., Zwerling, C. and Orav, E. J. (1992) 'Occupational risks associated with cigarette smoking in a prospective study', *American Journal of Public Health* 82(1) January.

Snook, S. M., Campanelli, R. A. and Hart, J. W. (1978) 'A study of three preventative approaches to low back injury', *Journal of Occupational Medicine* Vol. 16.

Society of Occupational Medicine (1992) 'What employers should know about HIV and AIDS'.

Stevenson, D. (1994) 'Repetitive strain injury: work-related upper limb disorders', 39 *Journal of the Law Society* 49.

Stubbs, D. A., Buckle, P. W., Hudson, M. P., Rivers, P. M. and Worringham, C. L. (1983) 'Back pain in the nursing profession: epidemiology and pilot methodology', *Ergonomics* 26.

Taylor, F. W. (1911) *Scientific Management*. New York: Harper & Row.

Terrence Higgins Trust (1993) *HIV/AIDS Positive Management*.

Tindall, A. (1993) *Tenosynovitis: A Case of Mistaken Identity*. Iron Trades Insurance Company Ltd.

UK NGO AIDS Consortium for the Third World (1989) *HIV/AIDS and Overseas Employment*. ECA Ltd.

Useful addresses

General

Confederation of British Industry
(CBI)
Centre Point
103 New Oxford Street
London WCIA 1DU
0171– 379 7400

Department of Employment
Caxton House
Tothill Street
London SW1H 9NA
0171– 273 6969

Department of Occupational
Medicine
University of Aberdeen
University Medical Buildings
Foresthill
Aberdeen AB9 2ZD
01224 685157

Faculty of Occupational Medicine
Royal College of Physicians
6 St Andrews Place
London NW1 4LE
0171– 486 2641

Health and Safety Executive
Baynards House
1 Chepstow Place
Westbourne Grove
London W2 4TF
0171– 221 0416
and through a network of offices
of the Employment Medical
Advisory Service (EMAS)

The Industrial Society
3 Carlton House Terrace
London SW1Y 5DG
0171–839 4300

Institute of Manpower Studies
Mantell Building
University of Sussex
Falmer
Brighton BN1 9RF
01273 686751

Institute of Occupational Medicine
University of Birmingham
Edgbaston
Birmingham B15 2TT
0121– 414 6022

Institute of Personnel and
Development (IPM)
Camp Road
London SW19 4UX
0181– 946 9100

Royal College of Nursing
20 Cavendish Square
London W1M 9AE
0171– 409 3333

Royal College of Physicians
11 St Andrews Place
London NW1 4LE
0171– 935 1174

Society of Occupational Medicine
6 St Andrews Place
London NW1 4LE
0171– 486 2641

Trades Union Congress (TUC)
Congress House
Great Russell Street
London WC1B 3LS
0171– 636 4030

Mental health

Alcoholics Anonymous
PO Box 1, Stonebow House
York YO1 2NT
0171– 352 3001 (London region
telephone service)

British Association for
Counselling
1 Regent Place
Rugby
Warwickshire CV21 2PJ
01788 578328

Lifeskills International Ltd
Wharfebank House
Ilkley Road
Otley LS21 3JP
01943 851140

Occupational Stress Indicator
Resource Systems
Claro Road
Claro Court
Harrogate HG1 4BA
01423 529529

Westminster Pastoral Foundation
23 Kensington Square
London W8 5HN
0171–937 6956

AIDS/HIV

Department of Health AIDS Unit
Friars House
157–168 Blackfriars Road
London SE8 8EU
0171–972 2000

Medical Advisory
Service for Travellers
Abroad (MASTA)
PO Box 14
Lee on Solent
Hants PO13 9LQ
01705 553933

Terrence Higgins Trust
52–54 Grays Inn Road
London WC1X 8JU
0171–831 0330

Smoking

Action on Smoking and Health
(ASH)
109 Gloucester Place
London W1H 3PH
0171–935 3519

ASH Northern Ireland
40 Eglantine Avenue
Belfast BT9 6DX
01232 663281

ASH Scottish Committee
8 Frederick Street
Edinburgh EH2 2HB
0131–225 4725

ASH in Wales
142 Whitchurch Road
Cardiff CF4 3NA
01222 614399

QUIT Ltd
102 Gloucester Place
London W1H 3DA
0171– 487 2858

Health promotion

CALM
PO Box 30

North District Office
Manchester M7 1NA
0161–428 5529
For computerised health assessment packages

Cancer Link
17 Britannia Street
London WC1X 9JN
0171– 833 2451

Health Education Authority
Hamilton House
Mabledon Place
London WC1H 9TX
0171– 413 1919

Health Education Board for
Scotland
Woodburn House
Canaan Lane
Edinburgh EH10 4SG
0131– 447 8044

Health Promotion Authority for
Wales
Brunel House (8th Floor)
2 Fitzalan Road
Cardiff CF2 1EB
01222 472472

Northern Ireland Health
Promotion Unit
The Beeches
12 Hampton Manor Drive
Belfast BT7 3EN
0232 644811

Tenovus Cancer Information
Centre
College Buildings
Courtney Road
Cardiff CF1 ISA
01222 497700
0800 526527 (freephone)

The Wellness Forum
Priory House
8 Battersea Park Road
London SW8 4BG
0171–222 2332
For information on health
promotion initiatives

Women's Nationwide Cancer
Control Campaign
Suna House
128 Curtain Road
London EC2A 3AR
0171–729 4688

Disabilities

Association of Disabled
Professionals
170 Benton Hill
Horbury
Wakefield
West Yorkshire WF4 5HW
01924 270335

Business in the Community
8 Stratton Street
London WIX 5FD
0171– 629 1600

Disability Matters Ltd
Berkeley House
West Tytherley
Wiltshire SP5 1NF
01794 341144

Employers' Forum on Disability
Nutmeg House
60 Gainsford Street
London SE1 2NY
0171– 403 3020

The Employment Service
Through a network of
Jobcentres for access to
Placing, Assessment and

Counselling Teams (PACTS)
and Regional Ability
Development Centres (ADCs)

National Council for Voluntary
Organisations
Regents Wharf
All Saints Street
London N1 9RL
0171– 713 6161

New Ways to Work
309 Upper Street
London N1 OPD
0171– 226 4026

Opportunities for People with
Disabilities
1 Bank Buildings
Princes Street
London EC2R 8EU
0171– 726 4961

Remploy Ltd
415 Edgware Road
London NW2 6LR
0181– 452 8020

Royal Association for Disability
and Rehabilitation (RADAR)
250 City Road
London EC1V 8AS
0171– 250 3222

Royal National Institute for the
Blind (RNIB)
224 Great Portland Street
London W1N 6AA
0171– 388 1266

Royal National Institute for Deaf
People
105 Gower Street
London WC1E 6AH
0171– 387 8033

The Royal Society for Mentally
Handicapped Children and Adults
(MENCAP)
MENCAP Pathway Employment
Service
MENCAP National Centre
123 Golden Lane
London EC1Y 0RT
0171– 454 0454

The RSI
Sheltered Employment and
Consultancy Services (SEPACS)
The Employment Service
Level 2
Courtwood House
c/o Rockingham House
123 West Street
Sheffield S1 4PQ
01742 596151

Training and Enterprise Councils
(TECs) in England and Wales,
and Local Enterprise Companies
(LECs) in Scotland. See your
local telephone directory for
details.

Women and employment

Daycare Trust
Wesley House
4 Wild Court
London WC1B 5AU
0171– 405 5617

Equal Opportunities Commission
Overseas House
Quay Street
Manchester M3 3HM
0161– 833 9244

Working Mothers Association

77 Holloway Road
London N7 8J7
0171– 700 5771

Employees and their environment

British School of Osteopathy
1–4 Suffolk Street
London SW1Y 4HG
0171–930 6093

British Standards Institution
2 Park Street

London W1A 2BS
0171– 629 9000
Ergonomics Society
Devonshire House
Devonshire Square
Loughborough LE11 3DW
01509 234904

HUSAT Research Institute
Loughborough University of
Technology
The Elms
Elm Grove
Loughborough LE11 1RG
01509 611088

Index